CREATING RAIN GARDENS

Cleo Woelfle-Erskine
and Apryl Uncapher

CREATING RAIN GARDENS

**Capturing the Rain for Your
Own Water-Efficient Garden**

TIMBER PRESS
Portland • London

PREVIOUS SPREAD:
Child's play in
a rain garden

The case study "Street Orchards for Community
Security" was adapted with permission from
desertharvesters.org and *Rainwater Harvesting
for Drylands and Beyond*, Vol. 2: *Water-
Harvesting Earthworks* by Brad Lancaster
(Rainsource Press, Tucson, 2008), available at
HarvestingRainwater.com.

Design by Jason Blackheart
Illustrations by Arthur Mount

Published in 2012 by
Timber Press, Inc.

The Haseltine Building
133 S.W. Second Avenue, Suite 450
Portland, Oregon 97204-3527
timberpress.com

2 The Quadrant
135 Salusbury Road
London NW6 6RJ
timberpress.co.uk

Printed in the United States of America

Library of Congress
Cataloging-in-Publication Data

Woelfle-Erskine, Cleo.
  Creating rain gardens: capturing the rain for
your own water-efficient garden/Cleo Woelfle-
Erskine and Apryl Uncapher.—Ist ed.
    p. cm.
  Includes bibliographical references and index.
  ISBN 978-I-60469-240-2
  I. Rain gardens. 2. Runoff—Management.
3. Ecological landscape design. I. Uncapher,
Apryl. II. Title. III. Title: Capturing the rain for
your own water-efficient garden.
  TD657.4.W64 2012
  635.9'5—dc23

                    2011032473

A catalog record for this book
is also available from the British Library.

CONTENTS

ACKNOWLEDGMENTS

JUST AS A RAIN GARDEN IS COLLABORATION AMONG RAIN, THE LANDSCAPE, AND LIVING CREATURES,

this book has been a rich and multifaceted collaboration. We worked together in every phase of research and writing and developed the methods presented here after conversations with rain gardeners across North America. We also field tested our site assessment and design methods on participants in hands-on workshops in Oakland, California, during 2010 and 2011.

Our shared love of gardening, photography, and ecological design and our love of water in all its natural forms run through the book. Rain, sun, clouds, soil, rivers, and plants are key collaborators in this effort as well.

Many of the site assessment and design activities in this book arose in collaboration with members of Greywater Action and the East Bay Academy of Young Scientists. Cleo's interest in sustainable water cultures evolved from conversations with members of the Environmental Justice Coalition for Water and the food justice movement and with Native Americans involved in restoration projects on the Nisqually, Klamath, Elwha, and Clark Fork Rivers. He hopes that this spirit of collaboration among people and across species comes through in these pages.

We would like to thank all of the rain gardeners who shared their rain gardens and offered expert review, especially David Hymel, Brad Lancaster, Brock Dolman, Jeanette Dorner, and Christopher Shein. Numerous rain guide authors and low-impact development experts in the cities of Seattle, Portland, Fort Wayne, Charleston, Berkeley, and San Francisco answered technical questions that improved the book's geographic scope. Nigel Dunnett contributed a U.K. perspective, case studies, and helpful comments on the manuscript. We take full responsibility for any remaining errors or omissions. Many thanks are due to Juree Sondker at Timber Press for proposing this book, Lisa DiDonato Brousseau and Eve Goodman for their help in editing, and the Timber Press crew for bringing it into the world.

Cleo thanks Joel Glanzberg for introducing him to integrative design and water harvesting traditions; Brock Dolman and Brad Lancaster for helping develop his ability to read the landscape and sky; Laura Allen and Greywater

Guerrillas/Action members past and present for plumbing and popular education collaborations; July Cole for watery explorations by trail, canoe, and pickup truck, philosophical contributions against Manifest Destiny, and the planning "flow" chart; Andrea del Moral for a decade of collaboration in art, politics, and water; and Jessica Diaz and Xarick Matsen for ongoing conversations on community science. He thanks Emilie Agosto and Anne Blair for South Carolina rain garden collaborations and shrimp and grits; his mother, Gretchen Woelfle, for manuscript review; his father, Peter Erskine, for photos of Oregon rain gardens; and both parents for their support of his gardening experiments past and present. He also thanks the U.S. Forest Service for a month of skywatching at Round Mountain Lookout near Mount Shasta, where he began work on this book.

Apryl thanks Bill Wilson, her greatest mentor in the field of environmental engineering and ecological design, for sharing years of experience and how nature knows best. She is indebted to her husband, Alex, for making every project better, and daughters, Nara and Reya, for keeping days fun and full of love. Apryl is grateful to her mother for inspiring a passion of nature and gardening, her father for encouraging answers through curiosity and commitment, and her sister for tirelessly offering faithful guidance. And finally, thank you to water heroes everywhere, creating a mindful water culture.

THE GARDEN IN THE RIVER, POOLS IN THE GARDEN

IN THE MIDDLE OF THE NIGHT,

RAIN BEGINS TO FALL. YOU HEAR IT ON the roof through your dreams. In the morning, clouds hang low over the city and rain falls lightly. Outside, rain cascades off your roof, bouncing down bell-shaped chimes suspended from the gutter, and splashes down on a large flat stone. The rain falls faster and rushes through your landscape, over stone and down grassy swales. The swales spill into shallow depressions filled with flowering shrubs and grasses. Water rises in the depressions, creating small pools throughout the garden. Just as the basins begin to overflow, the clouds thin and sun breaks through. Suddenly the tree over the rain garden is full of birds, and frogs sing from the shadows. Mist rises from stepping stones and leaves, and water in the basins sinks into the ground.

This is a rain garden: a simple depression in the ground that becomes a watery oasis every time it rains. Between storms, stone and wetland plants evoke flowing water, and sculptural rain chains mark the rain's passage from roof to ground. A rain garden recreates the prairie sloughs and woodland bogs that were filled in and paved over to build our cities and suburbs, creating a blooming oasis watered only by the rain. If every downspout led to a rain garden, much less water would run down streets and storm drains, and the flash floods that turn urban streams into raging rivers would end. Instead of washing contaminants from roofs, streets, and storm drains into the nearest lake or estuary, rain would sink into the soil, rehydrating parched landscapes and recharging aquifers. From a raindrop's perspective, our cities and suburbs would look more like natural landscapes.

Your house may lie far from the waterfront, yet your garden is in the pathway of a river (or lake, pond, or swamp) in the sense that water ending up in the river first flows through your garden. A river is not just a stream of water. All the roofs, streets, fields, and wildlands in the watershed drain into it. Any pesticide or fertilizer applied to gardens and every drop of oil or piece of dust that falls on roofs and streets sooner or later ends up in a body of water. Every water-shedding surface in the urban, suburban, or rural landscape can be a source of water pollution—or it can be a catchment surface for a beautiful and multifunctional rain garden.

A rain garden is a living water treatment system. It collects water rushing down a downspout or street gutter and slowly infiltrates that water into the ground. Because soil acts as a natural filter for pollutants, the water will be clean and cool by the time it ends up in the river, the lake, or the marsh nearby. In dense urban neighbor-

PREVIOUS SPREAD:
Many storm drains empty directly into a nearby stream or river.

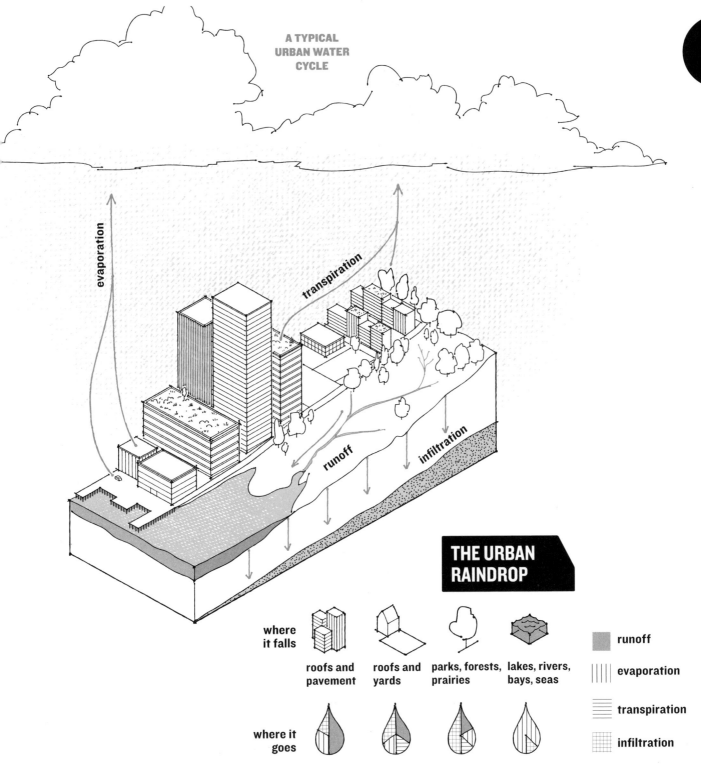

A TYPICAL
URBAN WATER
CYCLE

evaporation

transpiration

runoff

infiltration

**THE URBAN
RAINDROP**

where
it falls

roofs and
pavement

roofs and
yards

parks, forests,
prairies

lakes, rivers,
bays, seas

runoff

evaporation

transpiration

infiltration

where it
goes

Community members replace part of an asphalt playground with a rain garden.

hoods—or anywhere that space is at a premium—the concept of filtering runoff with plants and soil can be adapted to rooftops, vertical surfaces, and the soil under pavement. Collectively, these biological rainwater treatment strategies comprise the cutting-edge environmental field of low-impact development.

A home rain garden will collect rain that falls on your home watershed. If you live on the very top of a hill, your home watershed is exactly the size of your property. Otherwise, it's larger, because it includes rain that falls on higher neighboring land that then flows down onto your site.

This book will teach you how to welcome your home watershed's rainfall into a garden. We first explain how to map the boundaries of your home watershed and trace flows of water through the landscape. Next, we illustrate how to measure how much water falls on your home watershed in a typical storm and figure out how much of

that water you can capture and infiltrate in your rain garden. We present detailed discussions and instructions on digging, planting, and maintaining your rain garden, and we discuss other ways to catch and use rainwater in the garden, ranging from rain barrels to living roofs and walls to permeable pavement. We conclude with a chapter that shows how rain gardens can complement other watershed restoration strategies.

WHY PLANT A RAIN GARDEN?

There are three compelling reasons to build a rain garden. The first is practical: a rain garden allows you to conserve resources by working with nature rather than against it. The second is personal: a rain garden can be a visually interesting, low-maintenance feature of the home landscape, offering a dynamic, living waterscape that changes from day to day and season to season. The third is cultural: on the scale of a neighborhood or city, rain gardens can create new watershed relationships. Along with stream, lake, and wetland restoration and water conservation strategies like greywater reuse, rain gardens are a daily reminder of our dependence and influence on our local watery environments. Whether you are motivated by practical, personal, or communal concerns—or some combination of the three—this book is for you.

Our view of rain gardens gives rise to several biases and priorities, which we explain and develop throughout this book. We emphasize plants that are beautiful and multifunctional—they provide food, trellis or building material, passive home heating and cooling, and/or wildlife habitat. Rain gardens should lower a garden's water footprint and thus use no supplemental irrigation. We are avid do-it-yourselfers and explain how to make simple tools and hardware that you will need to build your rain garden. We mention special circumstances that require professional consultation, although you may consult a professional at any stage of the process. Rain gardens provide so many benefits for so little upfront and ongoing labor, they should be your primary rainwater harvesting strategy—supplemented by rain cisterns, permeable pavement, and living roofs and walls.

WHY RIVERS AND LAKES NEED RAIN GARDENS

By the late 1980s, when the term *rain garden* emerged in the United States, most people agreed that the rivers, lakes, and wetlands of the

WHAT'S POLLUTING OUR WATER?

THE U.S. CLEAN WATER ACT targeted sewers and factories, which are known as point sources of water pollution, because they discharge chemicals or sewage from a single pipe into a body of water. The treatment plants required by this legislation effectively reduced pollution from these sources, and thousands of lakes and streams became much cleaner as a result. These days, regulators are searching for ways to clean up the runoff from streets, rooftops, farms, and landscaped areas. This runoff is called nonpoint source pollution, and it adds up to a big water quality problem that is rarely regulated or even measured. For example, a recent Puget Sound Partnership study showed that 75 percent of pollutants in Puget Sound come from nonpoint sources. Oil, fertilizers, and pesticides that wash off urban and suburban landscapes and into nearby waterways compound problems caused by runoff from logging, mining, and agriculture. Rain gardens are an easy, economical, and sustainable way to reduce nonpoint source pollution.

industrialized world were in trouble. Dams and levees blocked and shifted rivers, changing the natural flood cycles that triggered salmon runs, frog egg-laying, and germination of riverside tree seeds. Sewers and factories spewed pollutants into rivers, lakes, and estuaries, poisoning aquatic animals and sickening people who fished or swam there. Rachel Carson's 1962 *Silent Spring* had captured a generation's imagination, and the regulations that followed, among them the U.S. Clean Water Act of 1972 and Australia's Clean Waters Act of 1970, reflected the concern that humans were poisoning the biosphere and destroying its capacity to sustain life.

During the second half of the twentieth century, cities and towns sprawled beyond their founders' wildest dreams. As roofs and pavement replaced forest, field, and prairie,

the amount of runoff increased as well, filling or even overtopping streams and flood-control channels that were designed for much smaller settlements. Where storm drains have been combined with sewers, storms now send raw sewage overflowing into rivers and estuaries. Contamination also causes beach closures and restricts fishing, threatening public health, quality of life, and local recreation-based economies. The motivation to reduce these regular floods of polluted water was partly legal, partly based on public health concerns, and partly cultural. Globally, water regulations increasingly require cities and counties to reduce trash, oil, raw sewage, pesticides, and other contaminants flowing into their waters. But how can you clean up pieces of trash and drops of oil spread across the urban land-

scape? Prevention is the best strategy, and trapping pollutants before they flow into waterways is the next best step. City planners turned to rain gardens to solve these two pressing problems caused by urbanization: pollution and flooding.

The idea of using vegetated depressions to infiltrate runoff has an earlier origin. In the 1970s, the field of ecological design and its Australian cousin, permaculture, emerged from the insight that ecosystems have an intrinsic value as well as many values to humans, among them natural resources, clean air and water, and climate regulation. A rain garden is an example of ecological design that works with nature to turn polluted rainwater (stormwater) into a beneficial resource. Across the industrialized world, rain gardens have proven integral to the restoration of lakes, rivers, wetlands, and estuaries by trapping pollutants and floodwaters, recharging aquifers, and creating wildlife habitat in cities and suburbs. Perhaps most importantly, home rain gardens serve as daily reminders of natural water cycles and how human activities affect local waterways.

Restoration projects in many large watersheds, among them Chesapeake Bay, the Everglades, Puget Sound, Lake Ontario, and the Murray-Darling Basin, include rain gardens, living roofs, permeable pavement, and other ecological design strategies to reduce pollution and flooding and to restore biodiverse green spaces to the urban environment. The rain garden is arguably the most impressive of these strategies, because a simple, vegetated basin provides so many environmental benefits, from wildlife habitat to passive cooling to pollutant removal.

You may have picked up this book because you are concerned about water pollution in a local creek, because you want to conserve water or create wildlife habitat in your garden, because you want a low-maintenance, environmentally sustainable landscape feature, or out of curiosity alone. Whatever your motivation or background, this book will explain in detail how to design, build, plan, and maintain the rain garden of your dreams; show you how to integrate the rain garden with other home water systems like rain barrels and living roofs; and encourage you to spread rain gardening through your neighborhood and community.

POOLS IN THE GARDEN

On the scale of a watershed, rain gardens can restore water flows through the urban landscape to a more natural, predevelopment pattern. Forests, prairies, shrublands, and swamps soak up much of the rain that falls, first on leaves, needles, and blades of grass, then in the duff layer of decomposing organic matter, and finally in soil pores and fungal webs. Water slowly drips from the plants to the ground and only runs off after a long rain. After the rain stops, water flows slowly downhill under the ground surface, reaching a lake or stream days or weeks after the storm. In contrast, rain that falls on roofs and streets runs quickly to the nearest storm drain, where it rushes to a lake or stream minutes or at most hours after the rain fell. By building pools throughout the city, planting their basins with plants that increase infiltration, and directing downspouts and gutters to these rain gardens, we can slow the flow of water through urbanized landscapes and trap pollutants in the soil, where they are much less harmful than they are in water.

Whether you use native or exotic plants in your rain garden, you are mimicking natural ecosystems that store, filter, and slowly release water to streams, lakes, and wetlands. Most rain gardens feature a mix of native plants and drought-tolerant non-natives. Native plants are a common choice for rain gardens because they are finely attuned to the local water cycle and require no fertilizers or pesticides.

CASE STUDY

PUBLIC GARDENS ON A FACTORY SITE by Nigel Dunnett

PLACE: Coventry, U.K.

DESIGNERS: Nigel Dunnett and Adrian Hallam

INSTALLED: Spring 2006

RAIN GARDEN SIZE: 3000 square feet (280 square meters)

ANNUAL RAINFALL: 27 inches (67 cm)

THE RAIN GARDENS on this commercial property are an excellent demonstration of how water-sensitive design principles can bring life to an otherwise sterile and lifeless site. Moreover, they illustrate that it is possible to use rain gardens to achieve large areas of very diverse planting in situations that would otherwise be out of bounds for greening.

Following a rebuilding program that resulted in an increase in the amount of car park areas and new buildings on the site, the factory managers noticed an increased flooding problem on site following heavy summer rainstorms. The increased areas of hard surface created a much greater amount of rainwater runoff, but the existing drainage system remained in place, and the pipes were no longer of sufficient diameter to handle the increased volumes of water. This overloading of the system caused runoff water to back up into the factory, with dirty water overflowing back into washbasins and bathroom areas.

The immediate proposal to deal with the problem was to dig up the existing drainage infrastructure and replace it with larger pipes, at great expense. At the time, our design team was working on a green roof for a building on site. We proposed the alternative approach of reducing the amount of rainwater runoff through rain garden installation, thereby enabling the existing pipes to remain. In so doing, the owners could create fantastic garden and recreation spaces for factory workers, at less cost than the traditional engineering approach of installing larger pipes.

Two large areas were created. First, a planting area 10 feet wide by 200 feet long (3 m by 60 m) was created alongside the factory cafeteria by removing the existing asphalt surface. The areas for the rain gardens were excavated to a depth of 3 to 4 feet (0.9 to 1.2 m). At the bottom of these basins, we placed a mix of crushed brick rubble and other free-draining material, and then the excavated areas were backfilled with an artificial free-draining substrate, consisting of 80 percent crushed brick and 20 percent municipal compost. An outdoor terrace deck was incorporated into the design. The downpipes from the adjacent building were disconnected so that runoff from the roofs discharged directly into the new rain gardens.

The second area was constructed over an existing concrete base. It was impossible to remove this concrete, so the rain garden feature (20 feet by 50 feet; 6 m by 15 m) was created as a large planter over the existing surface. The

The extensive planted areas, filled with perennials, replace part of a former parking lot and bring life to what would otherwise be a hard and sterile environment.

CONTINUED

AESTHETICS AND PRACTICAL CONSIDERATIONS

A rain garden can look like a formal flower garden, a wild streamside thicket, or an orchard. Rain gardens share design elements with Japanese and Chinese gardens—plants, rocks, stepping stones, urns, and natural-looking waterscapes—and can complement these styles of gardens. Rain gardens can also blend well with ponds or other water features by guiding water to the feature and then collecting and infiltrating the water that overflows.

Rain gardens range in size from bathtub-sized depressions to curbside or median planters that are the length of a city block. The size

Roof runoff flows through gutters, a downspout, and buried corrugated pipe to reach a rain garden.

of your rain garden will depend on how much rain you are trying to collect and how much space you want to devote to rain collection. If you have a large outdoor area, you can design a rain garden to infiltrate all of the rain that runs off the roof and paved areas such as patios, driveways, and sidewalks.

Once established, rain gardens are very low maintenance, as they require no irrigation and only occasional weeding. They differ from other gardens in two key respects. First, because rain gardens are designed to reduce stormwater pollution, they should never be fertilized or sprayed with pesticides or herbicides. Instead, use compost, companion planting, and biological pest control as needed. Second, a rain garden is more than just a planted basin. It includes the roof, gutters, downspouts, pipes, and channels that collect and transport water. This conveyance system that

same growing medium was mounded to a maximum depth of 18 inches (45 cm) and contained within an edging. Again, water was diverted from an existing roof.

In both cases, the rain gardens absorbed water from nearby roofs and the adjacent car park, but their construction also removed the equivalent areas of impervious surface from the sites. Because of the very free-draining nature of the growing medium and because the sites were surrounded by dark, reflective asphalt, the planting was adapted to withstanding warm periods with little rain, as well as periodic inundation. A naturalistic, meadow-like scheme was devised that provided a very long flowering season, combined with simple maintenance require-

ments: a single cut-back at the end of winter. The main plants in the gardens included *Aster* 'Purple Dome', *Stipa tenuissima*, *Miscanthus* 'Silver Feather', *Achillea* 'Moonshine', *Sisyrinchium striatum*, and *Allium schoenoprasum*.

The rain garden areas were an immediate success with both the factory workers and managers. The transformation from stark parking lot to exuberant garden was striking and instant. The rain gardens had a great impact on the working conditions in the factory, with a lot of interest being shown in the plants. The parking attendant, whose office sat at the entrance to the garden areas, was frequently asked to identify the plants because workers wanted to use them in their

own gardens. The factory managers were also able to gain publicity and marketing value from the installation because of its positive messages about sustainable practices, biodiversity, and water conservation.

It is highly unlikely—in fact virtually impossible—that the factory owner would have agreed to the replacement of around 3000 square feet (280 square meters) of car park within the complex with new gardens and high-quality planting purely for the amenity or beautification value alone. But by taking a more ecological approach, which also happened to be more cost-effective, the factory was able to solve the flooding and drainage problem and provide a fantastic resource for workers.

The rain gardens take water from the down-pipes of the building and surround a deck outside the workers' cafeteria.

moves rain from the roof to the rain garden basin is as important to consider as the basin itself.

ELEMENTS OF A RAIN GARDEN

Rain gardens go by many different names: infiltration basins, swales, bioswales, and earthworks. In this book, we refer to any depression that infiltrates rainwater as a basin. Long, skinny basins are swales, an Australian term popular in the field of permaculture. Swales that slope downward are diversion swales, and level ones are contour swales. As their names imply, diversion swales divert water toward a rain garden, whereas contour swales, which lie along level contour lines, allow water to pool and infiltrate into the soil. (Some water does infiltrate in diversion swales, too.) Earthworks is a blanket term for rain gardens, diversion swales, infiltration ponds, and other variations of stormwater detention built directly into the earth.

INFILTRATING WATER: BASINS AND SWALES

The classic rain garden is wide and roughly round, often with amoeba-like extensions. There's nothing wrong with straight lines in a rain garden, but simple rectangles and circles have much less edge than shapes with indentations or extensions. If your rain garden is in a rectangular courtyard or curbside strip, consider a more complex shape for the basin to create more visual interest. Complex shapes increase the length of the edge relative to the area of the basin, and edges and varied slopes create more microhabitats for plants and wildlife.

Rounded or rectangular basins are appropriate for sites that slope up to 15 percent. On steeper sites, however, it is important to

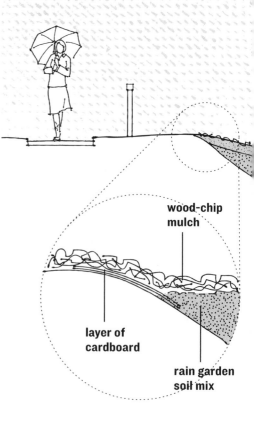

ANATOMY OF A RAIN GARDEN

wood-chip mulch

layer of cardboard

rain garden soil mix

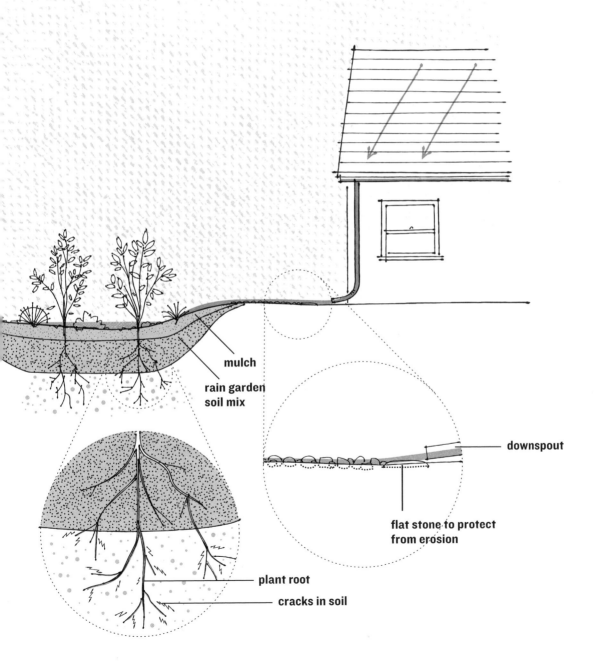

mulch

rain garden
soil mix

downspout

flat stone to protect
from erosion

plant root

cracks in soil

A flow-through
rain garden
planter captures
and treats runoff
in a small space.

spread water out across the landscape, because infiltrating a lot of water in one area (as classic round or rectangular basins do) can cause landslides. A contour swale—a long, skinny rain garden—is a better solution in this situation. Placing water-loving plants in the basin floor and fruit trees or shrubs along the raised berm will take advantage of the basin's varying growing zones and maximize soil stability. The roots will grow into the lens of water that moves slowly down the slope and infiltrates at the bottom of the basin.

MOVING WATER: RUNNELS, DIVERSION SWALES, AND BURIED PIPE

A diversion swale is a shallow channel that moves water from one location to another. It is similar to a contour swale, but its main function is to move water rather than infiltrating it. Diversion swales can be planted with grass or groundcovers or lined with cobble or river rock. A runnel is a diversion swale that carries water across hardscape (a patio, cement walkway, or driveway), and it can be made from wood, stone, metal, or concrete.

Diversion swales and runnels carry water across the surface of the landscape, which offers several advantages. These structures connect the rain garden visually to the roof or other catchment surface, highlighting the rain garden's stormwater treatment function to garden visitors. Surface structures are easy to inspect for debris buildup and easy to clean out when clogged.

Rain can also flow invisibly to the rain garden, through buried drainpipe the size of your downspout or larger. Drainpipe can carry water under walkways or lawns and down steeper slopes than surface channels can. Rigid PVC and flexible corrugated drainpipe are both appropriate for use in rain gardens.

CATCHING RAIN: ROOFS, DRIVEWAYS, AND OTHER SURFACES

You may not think of the roof as part of your home landscape, but it is an essential part of a rain garden. The catchment includes any surface that sheds rain, including a roof, driveway, patio, or lawn. Rain gardens can also infiltrate runoff from streets and parking lots. Any type of roof or paved surface can catch rain for a rain garden, but don't use runoff from parking lots or streets to grow edible plants in your rain garden, because plants may take up heavy metals from such highly contaminated runoff.

If your house has gutters, you can use either a downspout or rain chain to move water from the roof to the diversion swale or drainpipe. Rain chains also work with canales, spouts that direct water off of the flat-roofed houses common in the American Southwest. If your house doesn't have gutters, you can collect rain in a diversion swale at the base of the eaves.

LAST WORDS

BEFORE WE HONE in on technical details for rain garden construction, we offer a set of activities that will give you a new perspective on your home watershed and the natural areas nearby. We invite you to inspect your downspouts during a rainstorm, seek out the natural and cultural history of your neighborhood, and make maps and sketches to get a better understanding of water flow through your site. While none of these activities are strictly necessary to make a rain garden, they will help you design a beautiful and functional rain garden that perfectly suits your site conditions and goals. Try the ones that strike your fancy, and skip the rest. We bet you'll never see water the same way again.

A CHOOSE-YOUR-OWN-ADVENTURE RAIN GARDEN PLANNING GUIDE

WATER, SOIL, ROCK, AND PLANTS

CAN COMBINE IN SO MANY WAYS THAT

it can be daunting to design the rain garden of your dreams. What's more, figuring out how your rain garden will function poses its own challenges. But before you worry about calculating the runoff from the roof or rerouting downspouts, it's good to think in the most general terms about what you want your rain garden to do, how you want it to look, and how you will interact with it in different seasons.

This chapter will help you figure out what kind of rain garden best fits your gardening style and your site's prospects and limits. We begin with questions that will help reveal what kind of rain garden you want. Then we guide you in exploring your home waterscape. We'll describe the best sites for a rain garden—and where one should not go.

Along the way, we'll send you on optional adventures that won't fit between the pages of this book: finding natural and constructed rain gardens in your area and learning more about how human activity has changed the flow of water through your local watershed. A visit to a demonstration garden, natural history museum, or flowering roadside ditch can spark the imagination at any stage of rain garden design. These adventures can also give you a greater understanding of where you live, introduce you to local watershed stewards, and inspire other watery endeavors.

Rain garden design can be fairly lackadaisical or precise and involved, depending on whether you're more comfortable with a rule of thumb or a calculator. If you shy away from math, we recommend that you take the rule-of-thumb approach presented in the main text. You may end up with a rain garden that is slightly larger than necessary, but it will be a beautiful and functional system nonetheless. If you're comfortable with basic geometry and confident with a calculator, begin with the rule-of-thumb method and then check your design with the Engineer's calculations presented separately from the main text. Either way, don't skip the goals or site assessment exercises outlined in this chapter—both are necessary for rain garden success.

We've divided the planning process into three parts: considering your rain garden goals and dreams, exploring your home watershed, and assessing potential sites for your garden. Design is not really a linear process, and we like to think of it as a looping one. A visit to a demonstration

PREVIOUS SPREAD: Marina Wynton's goals for her rain garden included infiltrating the runoff from most storms and showcasing native Pacific Northwest plants.

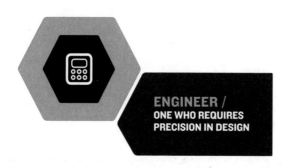

ENGINEER /
ONE WHO REQUIRES PRECISION IN DESIGN

WHAT KIND OF RAIN GARDEN DO I WANT?

Questions to ask yourself	What is my local watershed's history?	What natural rain gardens grow here?	What are my goals?	What is possible for my site?
Goals you hope to achieve	food	color	biodiversity	wildlife habitat	water conservation
Your current site	paved desert poor drainage		level site filled marsh		cleared forest sloping site
What was there before	desert spring		cypress swamp		willow thicket

Your ideal rain garden

ephemeral pool   restored marsh   woodland glade

Brussels
sprouts thrive
in this rain
garden.

garden or nursery might elicit new plant possibilities, while assessing your site might reveal a drainage problem that the rain garden could address. If you're intimidated by technical details, you can read about dreaming up your garden first, then dip into the mapping exercises once you have your dream garden in mind. We encourage you to cycle through the assessment, exploration, and dreaming up sections until you have a clear image of the rain garden you want. During the design process, you will answer questions such as: What is my local watershed's history? What natural rain gardens grow here? What's possible for my site? What are my goals for the rain garden?

RAIN GARDEN GOALS AND DREAMS

Why do you want a rain garden? How much time and money do you have to build and maintain it? What elements are missing from your home landscape that a rain garden might provide? Considering these questions is a great way to start planning your rain garden. When you think about your rain garden, think about how you want to use it. Do you enjoy cooking, eating, or simply sitting outside? What views do you want from the house? Can you integrate your neighbors' landscapes into your garden?

Rainwater harvesting systems—including rain gardens—vary enormously. Their designs depend on budget, maintenance commitment, aesthetics, and other goals like food production or wildlife habitat. One system might suit these circumstances:

> *I want to maximize animal habitat in a large yard. I'm not much of a gardener, but I want to do my part for the river by storing all runoff from my roof and driveway on site. I own the house and can buy any materials or plants I need, but I won't have much time for maintenance.*

ENGAGE YOUR ARTISTIC SIDE

AS YOU DAYDREAM about why you want a rain garden, try expressing your desires through art. Get out some paper and colored pencils or crayons and draw your dream garden, or collage one together from magazine photographs. This may give you insight into how wild, manicured, colorful, or subdued you want your rain garden to be.

31

Rain gardens can be planted with species that welcome insects, such as this painted lady butterfly (*Vanessa cardui*).

CASE STUDY

A ROADSIDE DITCH BECOMES A BLOOMING OASIS by Rick and Lucy Briley

PLACE: Fort Wayne, Indiana, U.S.A.

DESIGNERS: Rick and Lucy Briley

INSTALLED: Spring 2009

RAIN GARDEN SIZE: 2000 square feet (185 square meters)

ANNUAL RAINFALL: 36 inches (90 cm)

underground service lines and utilities.

A rented rototiller was a lifesaver. We used it to loosen the compacted clay soil and to till in soil amendments. The depth of the garden varies from 6 to 12 inches (15 to 30 cm). We designed the rain garden to be higher on one end and added some large rocks and ornamen-

The rain garden retains the runoff from heavy storms and adds beauty to the landscape.

OUR HOME IS LOCAT-ED toward the bottom of a street with a slight incline, but during heavy rains you'd have thought we lived on a river. The rainwater traveled next to the street in a natural depression, and as it reached our property it flowed over our driveway and into our yard on its way to the sewer grate. We read about rain gardens in the local newspaper and began to hunt for more information. It didn't take long for us to realize that a rain garden could take pressure off the sewer system and make our front yard more attractive.

We attended a workshop hosted by the City Utilities Department, and within a couple weeks we were excavating our rain garden. The workshop helped tremendously by explaining soil type, infiltration rate, size, depth, layout, and how to improve soil. It also helped us get city permission to construct the rain garden on the easement and figure out who to call about locating

CONTINUED

QUESTIONS TO SPARK THE IMAGINATION

▶ What type of natural feature can your rain garden mimic?

▶ Which rain gardens in this book or your local area do you like?

▶ How can your rain garden enhance people's experience of your garden or house? Can it provide shade, birdsong, evaporative cooling, flowers?

▶ What would you love to look at?

▶ What would entice you out of the house and into the garden on a summer evening?

▶ What would enhance your walk to the front door?

▶ What plants from your childhood do you miss?

▶ What colors excite you?

▶ What wildlife might you like to provide habitat for?

Another person may have these goals, requiring a different kind of design:

My two young children and I keep a big orchard. I want to stop using the hose to water it. I want a system all three of us can help build and use safely. I don't want to spend more than $100 on the rain garden, but we have lots of friends to help us dig.

Yet another system would work within these constraints:

I rent a house in a dense urban neighborhood and have a tiny, cement-covered space with little sun. I want to create an oasis that needs no irrigation.

Try to frame your own goals (such as habitat), requirements (safety or simplicity), and limits (low maintenance) in this way.

To spark ideas of rain garden goals, picture water as it flows through your site—in pipes through buildings, over roofs, across lawns, down streets, and into sewers. Then jot down answers to these questions:

▶ **What is rain doing on the site now?**

▶ **What would you like it to do—irrigate fruit trees, grow a blooming oasis, fill a birdbath, or cool a patio?**

▶ **Besides catching water, what else can a swale or rain garden do? For example, can the berm of a swale also be a raised path? Can tall plants block street dust and noise?**

▶ **Are there places in your landscape where infiltrating more water would cause problems?**

Identify your top priority as you consider your goals. Is it to feature particular plants (edibles, natives, dense shrubs, plants that attract butterflies or have purple flowers) or a style of garden (formal or wild)? Less obvious goals might include removing lawn, dealing with a boggy or

tal elements. Next we planted species native to Indiana: purple aster (*Aster patens*), candy mountain foxglove (*Digitalis purpurea*), pink *Astilbe*, purple coneflower (*Echinacea purpurea*), great blue lobelia (*Lobelia siphilitica*), butterfly weed (*Asclepias tuberosa*), and wild senna (*Senna hebecarpa*). All of these plants have really thrived. After planting, we put down about 2 inches (5 cm) of coarse hardwood mulch.

We're now going on our third year as rain gardeners. The rain garden has been a great lesson in improving our environment while growing a beautiful flower garden. We've noticed that during a heavy rain, water starts to flow over the driveway but then it gets absorbed into the rain garden. It has to rain several inches quickly before we see water spilling over the rocky berm at the downward end of the garden. Like most gardens, a rain garden may always be a work in progress. Don't be afraid to try out new ideas until you find the right mix for your own rain garden.

Homeowners Rick and Lucy Briley with their rain garden

shady spot where nothing grows, reducing how much you need to water and fertilize your landscape, providing play opportunities for kids, and growing a privacy screen.

THE AESTHETICS OF YOUR DREAM RAIN GARDEN

Most people have a dream house, dream car, or dream workshop, and all gardeners have a dream garden. Although you may not know it, you also have a dream rain garden—it's just a matter of dreaming it up. Here we're concerned not with the functional but with the aesthetic experience of the rain garden.

Rain gardens delight with bright colors, darting dragonflies, and birds and a variety of textures, colors, scents, and sounds. Logs, stones, and mulch provide habitat for lizards, frogs, salamanders, and bugs—and an irresistible playground for kids and grown-up naturalists. Rain gardens can be designed to provide forage and cover for rare and spectacular butterflies and birds. The scents of flowers and herbs evoke moods and memories, especially after a rain. The sounds of buzzing bees, rustling leaves, and rain falling on stone can be energizing or tranquil. When dreaming up your rain garden, consider not only how it will look, but how it will sound, smell, and feel under your hands or bare feet.

You may start dreaming up your rain garden by considering what would entice you into the garden on a summer evening. Do you want to walk barefoot over fragrant chamomile or round river rock? Would you like to hear tree frogs calling, cicadas rasping, water splashing down on stone? What colors of flowers or berries draw your eye?

You can also pick a theme for your rain garden. Maybe you want to attract butterflies or birds or provide habitat for frogs, salamanders, or newts. Find out what types of plants your featured animal needs for food or shelter and what other elements—such as logs, stones, bare soil, or a small pool of water—that it needs. Then get out the art supplies and make a sketch or collage of how these elements might fit to-

gether. If you live in the city but miss the country, you could make a bower of mountain azalea the backdrop for your rain garden. If you're an uprooted urbanite who misses botanical gardens, your rain garden could feature species from around the world. If you love edible berries or blue flowers or carnivorous plants or even rock gardens, start to imagine a rain garden organized around that theme. Once you understand your site and the needs of rain garden plants, you can visit a butterfly sanctuary or edible plant nursery and find species that can thrive in some part of a rain garden.

Grasses and rushes submerged by high river flows would thrive in the wet areas of rain gardens.

Also consider how your family and friends will inhabit the garden. Your rain garden can be a basin surrounded by lawn, or it can wrap around a covered gazebo so that you can sit outside during a storm and watch it fill with rain. Tall stepping-stones could let you cross the basin just above the water after a storm. Water wheels, toy boats, or even a rubber duck can make the rain garden your child's favorite place in the yard. Put a rain garden along the path to your front door and you'll be treated to a daily riot of lizards, birds, and dragonflies.

EXPLORING YOUR HOME WATERSHED

A rain garden mimics natural ecosystems that catch storm runoff and slowly release it into the soil and ultimately into streams, lakes, or wetlands. Depending on where you live, this natural ecosystem may be a prairie slough, an old growth forest, a cypress swamp, or a desert

RESOURCE

To find out more about your local stream and its advocates or to adopt your watershed, check out epa.gov/adopt (in the United States) or http://www.rivernet.org/welcome.htm (worldwide). You can find a simple identification guide for stream critters, suitable for use in Australian, North American, or European streams, at www.roaringfork.org/images/other/aquaticinvertebratesheet.pdf.

wash. Because rain garden plants should withstand local variations in precipitation, native species are an obvious first choice. For inspiration, look to your local creek, river, pond, bog, or lakeshore.

But what if your local waterway is hidden? If the salt marsh was filled to make your neighborhood, if the floodplain is cut off from the river by levees, if the lake is surrounded by riprap, or if the stream flows through a storm drain under the street, don't despair. Begin your watershed exploration with natural history books and websites, native plant societies, and tribal museums. Whether your area was prairie, forest, marsh, floodplain, or sand dune, consider how water flowed through the landscape. Did rain land on pine needles or oak leaves, then drip down onto a thick layer of duff? Did runoff flow down blades of grass and sink into soil packed with grass and wildflower roots? Was the soil below well-drained sand or waterlogged peat in a bog or marsh? Understanding your site's history will help you choose the best rain garden elements to recreate a natural and healthy ecosystem and make the most of the rain.

GETTING YOUR FEET WET: LOCAL WATERSHED EXPLORATION

The next time it rains, splash down to your local waterfront. If you find trash and oil-slicked water, you're not alone. According to the Environmental Protection Agency, half of all U.S. waterways are unswimmable or can't support fish and wildlife. Only five of Europe's fifty-five major rivers are considered pristine by the United Nations Environment Programme, and one-third of Australia's rivers are listed as impaired in the Australian Natural Resource Atlas. The most common source of water pollution in urban and suburban areas is roof and street runoff, which is why some local governments in North America, Europe, and Australia are promoting rain gardens.

Start at the headwaters or the highest point you can find and walk downstream. If you have

OPPOSITE:
Rain gardens can include food harvesting themes, such as this garden featuring ripe squash.

to scramble through dense thickets, you're in luck. Willows and other streamside vegetation cool streams and provide crucial habitat for birds and mammals. Make a note of what species you see in abundance (or sketch the plants and do some research at local nurseries or botanical gardens to identify them). Once you emerge, maybe you'll find beaver dams, which cool streams, create wetlands, and shelter fish. Look for frogs, which are sensitive creatures that have low tolerance for polluted environments. Do you see wildlife you would like to attract to your garden?

If you see bare, muddy banks, look for cows or new development, which increase runoff and erosion by removing plants that stabilize stream banks. Erosion adds pollutants to streams and clogs fish spawning grounds with mud. Make notes or sketches of what kind of plants are colonizing the bare banks. These are hardy species adapted to steep, eroding soils and may be good choices for addressing erosion issues on your site.

Many fish, as well as swallows, lizards, and frogs, eat water bugs. Insect larvae clinging to wet rocks provide clues to water quality. Stoneflies need pristine water, whereas rat-tailed maggots survive in scummy stormwater. Ask fishers what they're catching. If you see no bugs or fish, consider joining or starting a local group to help restore your watershed.

If you can't find a stream nearby, it may be buried in a storm drain. Contact your local creek group for a storm drain map, then trace the storm drain (aboveground!) through the city streets. Can you picture a willowy stream flowing there?

As you explore your watershed, seek out hillsides and floodplain areas that host a diverse mix of plants. If your watershed no longer boasts healthy streams or slopes, visit a nearby natural area instead. You can use these areas as study sites for your rain garden. Bring a field guide to native plants, a trowel, and a small notebook.

Choose a trail that passes through varied terrain—up hills, past ponds or bogs, and along creeks or shorelines. Notice where water flows and pools. What patterns does water make as it flows down a dirt trail? Where does water collect into rills and channels, and how steeply does the ground slope there? Notice signs of erosion—bare soil, crumbling slopes, gullies, and muddy areas. Pay particular attention to places where the water flows slowly downhill without getting muddy. You can mimic these natural sloughs in your landscape by building diversion swales to move rainwater from the downspout to a rain garden. On your walk, look for depressions that catch and hold water. With your trowel, dig down into the soil. Is it sandy, mucky, clayey? If you can, come back when it hasn't rained for a couple of days and see whether the basin still holds water. If it does, there's probably clay close to the surface.

Notice what plants are growing in and near natural depressions. They are contenders for a place in your rain garden. You might see sedges, rushes, grasses, and wildflowers or shrubs like dogwood and willow. Do the plants grow in distinct zones based on how deeply they're submerged? Note the plants you see and whether they're growing in the sun or shade and how much water they prefer. Do you see insects or birds on or around any plants? Think about which plants you might like in your rain garden because of their color, shape, flowers, or edibility or for wildlife habitat.

HISTORICAL HYDROLOGY: YOUR NEIGHBORHOOD BEFORE IT WAS BUILT

Long-time residents may remember how urbanization changed water flows through your watershed. They may have fished at local creeks before they were buried in storm drains. They may remember fish spawning in streams that now harbor only algae, or birds visiting wetlands now buried under strip malls. When you meet such people, ask them about changes in the landscape. Have creeks dried up or turned into muddy gullies? Has a change in land use from forest to farmland or

OPPOSITE:
This rain garden under a mature Douglas fir lies over about a quarter of the tree's roots. Because the soil is well drained, the roots stay relatively dry.

farmland to city increased flooding?

A powerful tool for finding your place in your watershed is Google Earth. You can download the program for free from the internet. Once you find your house, explore the territory in satellite view. The tilt function gives a three-dimensional view. Notice how much of your watershed is developed and paved. Every roof and road is a source of pollution and a potential catchment area for a rain garden. Look at where your house is in relation to creeks, lakes, ponds, wetlands, and springs. Then switch to terrain view and trace your local creek up to the top of the watershed (contour lines get closer together and V upstream along waterways).

To see how your watershed has changed over time, use Google Earth's history slider. The red lines on the slider represent the dates of aerial photographs, which often go back to the 1930s. By moving the slider forward in time, you may be able to see your neighborhood being built. As you plan and build your rain garden, you will reconstruct the living systems that trapped water and nutrients before developers scraped the topsoil bare and paved the land.

ASSESSING YOUR SITE

Your site assessment can be as low-tech as making a rough sketch of water flow on a base map or

it can be a meticulous plan. Here we explain how to make a detailed assessment of your rain garden potential by mapping water flow, slope, infrastructure and vegetation, and soil types on transparencies or tracing paper, and then layering these over a base map. The base map can be an existing garden or site plan or an aerial photograph printed from Google Earth.

You can also use Google Earth to get a rough estimate of your roof area and potential rain garden sites. The free version of Google Earth includes a ruler tool, and the units can be set to feet, meters, kilometers, or miles. Or you can measure roofs and pavement directly using a tape measure. Measure the length and width of your roof as well as other hard catchment areas such as driveways and patios, and then calculate each area by multiplying the length times the width. Record the area of all hard catchment surfaces on the worksheet at the end of this chapter. You will use this area in the next chapter to calculate how much rain you can collect from your site.

MAPPING WATER FLOW

The paths followed by water as it flows through your home landscape will determine where you can most effectively place your rain garden to catch the most water and how large the garden will need to be. First, go outside and look at your house. Do you have gutters? Where are the downspouts? What areas of your yard are lower than the house? The headwaters of your home watershed are at the top of your roof. Mark this spot (and the top of any other roofs, including your neighbors') with a star on the base map. Now imagine rain falling on the roof. Trace its path to the gutters and down the downspouts. Draw arrows on the base map showing the water's flow, and mark each downspout location with an x.

Next, consider each downspout in turn. When the water splashes down, where does it go? If it hits a sidewalk or driveway, notice what direction the ground slopes and mark the places where you think water will flow with arrows. Draw dotted lines around depressions or vegetated areas where water

ENGINEERS / You can use a separate transparency to map each site characteristic—water flow, slope, infrastructure, and soil—that will affect rain garden placement. The stacked transparencies will make it easy to see the best sites for your rain garden

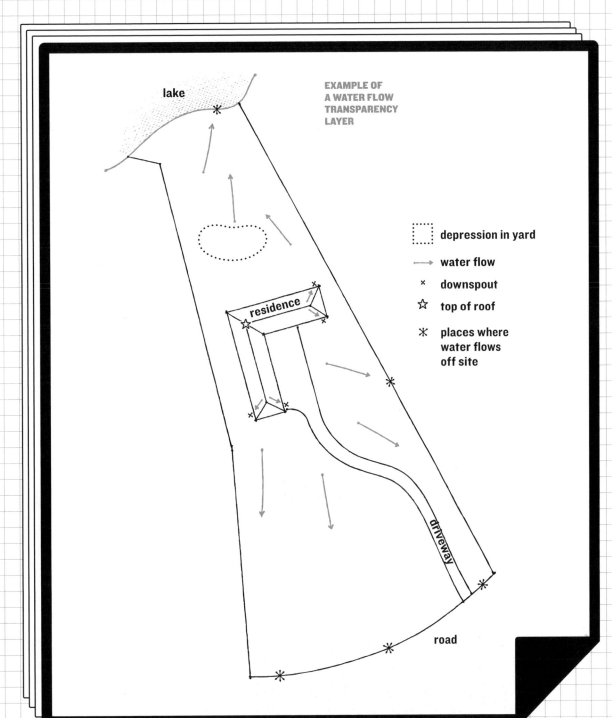

lake

EXAMPLE OF
A WATER FLOW
TRANSPARENCY
LAYER

⬚ depression in yard

→ water flow

× downspout

☆ top of roof

✳ places where
water flows
off site

residence

driveway

road

sinks into the soil. Continue walking around your house, marking runoff from steps, patios, sidewalks, and driveways. Finally, mark the places where water leaves your site with asterisks.

The next time it rains, test your prediction of water flow through your home watershed. Repeat your survey and add or remove arrows based on your observations. (This is also a good time to make sure all of your gutters are working properly.) Estimate and note what percentage of the water flows out of each downspout. In theory, the flow from a gutter should be proportional to the area of roof it drains. In practice, gutters usually slope toward one end and send more water down that downspout. If rain is not in your forecast, you can mimic what will happen by turning on a hose and letting it run at the bottom of each downspout. Then trace the path of water flow, and change your map if necessary.

ESTIMATING SLOPE

If your site is steep or at the top of a hill, you need to make sure that your rain garden won't cause a landslide. If your site is flat, you need to check that overflow from a rain garden basin won't run back toward any building foundations.

You can estimate slopes visually or measure slope with a laser level, string level, or, our favorite tool, the water level. (See the box "How to Build a Simple Water Level" in chapter 3.) Slope can be measured as a percent, in degrees, or as a ratio. We think that percent slope is the most intuitive and use this measurement throughout the book.

Consider the drawings of triangles. In each case the vertical rise represents 100 feet (or meters)—the units are not important—and the horizontal run varies from 100 to 400 feet (or meters). The percent slope is the rise divided by run times 100. So, the vertical rise over a 100-foot (or 100-m) run is the percent slope. In general, the percent slope of a site can be classified into four grades:

▸ **Gentle: less than 5 percent**
▸ **Moderate: 5 to 15 percent**
▸ **Steep: 16 to 30 percent**
▸ **Extreme: greater than 30 percent**

A VARIETY OF SLOPES IN PROFILE

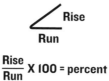

Rise
Run

$$\frac{\text{Rise}}{\text{Run}} \times 100 = \text{percent}$$

1 : 1 = 100% = 45°

2 : 1 = 50% = 26°

3 : 1 = 33% = 18°

4 : 1 = 25% = 14°

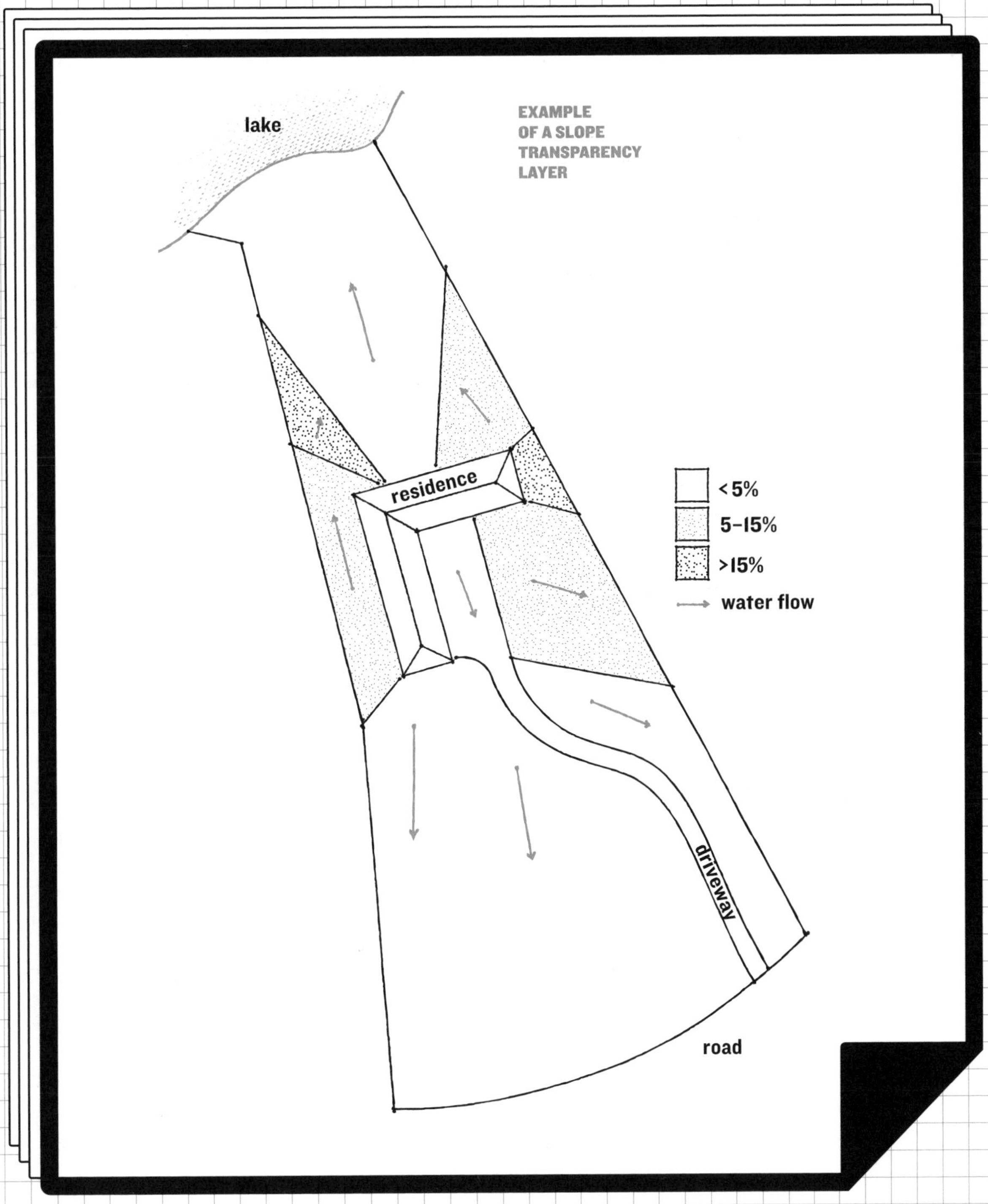

EXAMPLE
OF A SLOPE
TRANSPARENCY
LAYER

lake

residence

driveway

road

☐ < 5%

▨ 5–15%

▨ >15%

→ water flow

You may be more used to thinking of slope in ratios or degrees. For a site with a 1:1 slope (representing run:rise), the ground rises 100 feet (or m) for every 100-foot (or 100-m) run. The slope represents the angle where the diagonal of the triangle (the hypotenuse) meets the run, in this case 45°. A 2:1 slope would rise 50 feet (or m) over a 100-foot (100 m) run, and the slope would be 26°.

Measuring slope is a fairly simple task. You'll need a partner, a level, and two measuring tapes. Stand at the top of the slope, at the point where the slope gets steeper. Have your partner stand at the bottom of the slope, where it flattens out. Stretch a tape measure between you and hold it level (use the level to check). Record the horizontal distance between you. Then measure from the ground to the height of the horizontal measuring tape. Divide the vertical distance (in feet or meters) by the horizontal distance to get the percent slope.

Map the slopes on your property by first dividing the terrain by changes in slope. Draw a line on the map wherever you see a change in the steepness of the site (needless to say, this is somewhat subjective), then connect the lines to make boxes around areas of similar slope. Draw arrows inside the boxes to indicate which direction the land slopes. Then estimate whether the slope in between each line is gentle, moderate, or steep using the measuring technique described above. Stipple moderate slopes with dots, and leave gentle slopes blank. Fill steep slopes with a darker stipple or a solid color; these are the least desirable rain garden sites, because infiltrating too much water can cause a landslide. Likewise, color areas that are higher than the bottom of the downspouts a solid color, because water would have to be pumped uphill to reach a rain garden in such areas.

It's simple to make rain gardens on sites that slope up to 15 percent, so try to locate your rain garden on gentle or moderate slopes. Water won't flow across flat ground or uphill (or through level or upward-sloping pipes), so make sure that the ground slopes at least 2

percent between the downspout and rain garden site. If the ground is flat, you can move water to the rain garden through a pipe that slopes downhill at a 2 percent grade—about 0.25 inch/foot (2 cm/m). Be sure to place the rain garden close to the downspout or it will end up very deep in the ground.

You can build a rain garden on slopes between 15 and 30 percent with proper design and guidance. In this case, consider building contour swales or terraced rain garden cells connected by pipes or cascading overflow channels. Consult an expert for help designing a rain garden on any slope you've estimated to be more than 15 percent. Infiltrating more rainwater on these sites could cause the whole slope to fail and slide downhill.

MAPPING INFRASTRUCTURE AND VEGETATION

Next, draw a dashed line on the map to show a buffer zone around building foundations and septic leach fields, where a rain garden could cause flooding or mold problems. Rain garden basins require a buffer of 10 feet (3 m) from foundations and leach fields. Some rain garden designers will plant as close as 6 feet (1.8 m) from a foundation if the ground slopes away from the building and the rain garden has a bomb-proof overflow. If you have a basement or have witnessed flooding on your property, be conservative and keep rain gardens at least 20 feet (6 m) away from foundations. Because rain gardens infiltrate water into the soil below them, they can flood septic leach fields and make septic effluent rise to the surface—so definitely mark leach fields as sites to avoid. Utility lines should also be noted on the map.

You can map vegetation on the same transparency or use a different one. Mark natural depressions where runoff collects and drains well with a dashed line. Don't dig up these natural rain gardens. Also mark existing garden beds and large trees with dashed lines. Tree roots extend beyond the edge of the canopy, and digging through their roots could dam-

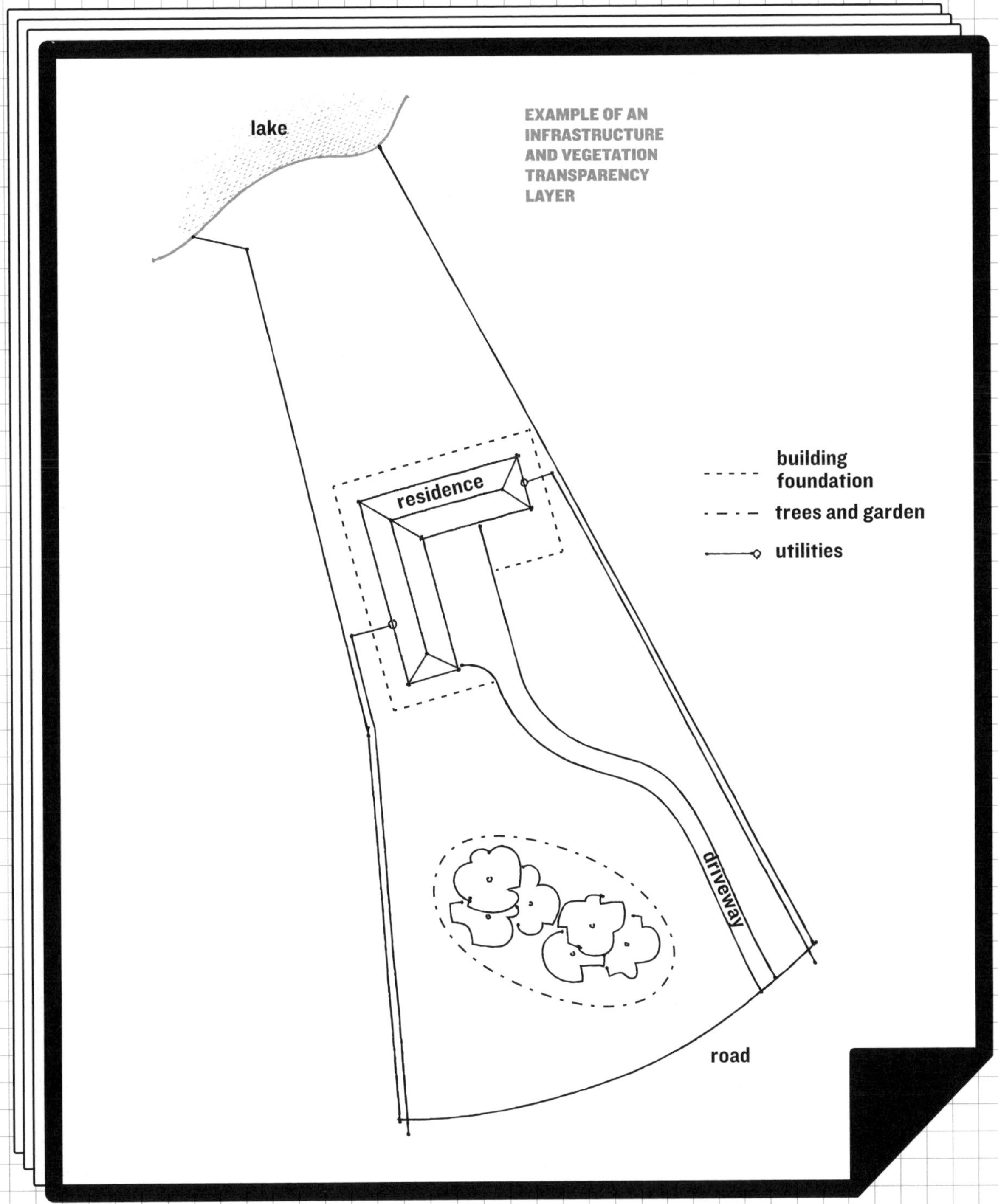

EXAMPLE OF AN INFRASTRUCTURE AND VEGETATION TRANSPARENCY LAYER

lake

residence

driveway

road

- - - - building foundation

- · - · - trees and garden

——○ utilities

age them. Consult an arborist if you have any concerns. In addition, it's often difficult to establish other plantings in shady areas under established trees. You may be able to dig a shallow rain garden without interfering with the tree roots, or build a berm on top of the soil. Trees that tolerate periodic flooding include Sitka dwarf alder (*Alnus viridis* ssp. *sinuata*), Japanese maple (*Acer palmatum*), vine maple (*Acer circinatum*), mountain ash (*Sorbus*), and willows (*Salix*).

Layering the water flow, slope, and infrastructure and vegetation transparencies on top of each other will reveal the best places to locate a rain garden. If you have lots of choices, you're in luck and can make your choice based on optimal aesthetic placement or include more than one basin. If your overlays show few or no good sites, you may have to make concessions and place a rain garden in a less-than-ideal location.

CHOOSING POTENTIAL SITES

Once you have identified potential rain garden sites, review your site map with the various overlays and ask yourself these questions: Does my water harvesting design use gravity to move water or fight gravity? Can I avoid using a pump? How can my rain garden design account for future plantings and structures? How can my water harvesting system be flexible, expandable, and able to be rerouted?

When choosing a site for your rain garden, *avoid* sites that are:

▶ Less than 10 feet (3 m) from building foundations and septic leach fields
▶ Where the water table is shallow, that is, less than 1 foot (30 cm) from the bottom of the rain garden
▶ In poorly draining depressions (these spots are okay for a pond, bog, or vernal pool garden)
▶ Over utility lines

▶ Under trees that don't tolerate flooded soil
▶ Under mature trees, where roots will limit rain garden size and make digging more difficult
▶ On slopes greater than 15 percent (or up to 30 percent with expert consultation)
▶ In locations that are higher than the bottom of the downspout—you don't want to have to use a pump to move water (in chapter 6 we discuss how using a cistern may allow you to build a rain garden slightly uphill from the downspout).

TESTING SOIL DRAINAGE

Knowing how quickly your soil drains is another key piece of the design puzzle. As a gardener, you probably already know a lot about your soil's fertility, but you may not have thought much about drainage. Most rain garden plants are much less picky than edibles or ornamentals when it comes to soil nutrients, but they need particular drainage conditions. Also, how quickly the soil drains will influence how large to make your rain garden.

Most rain gardens are excavated to a depth of 9 to 24 inches (22.5 to 60 cm), shallower in well-draining soil and deeper in slow-draining soil. If you know your soil drains well, dig a 1-foot (30-cm) test hole. If it drains slowly, dig a 2-foot (60-cm) hole.

Dig a test hole at each of the potential rain garden sites. Look at the soil in the hole. The top layer will probably look darker because it contains more organic matter. Beneath the topsoil, the subsoil will be some mix of sand, silt, and clay. Is the soil at the bottom of the hole topsoil or subsoil? Examine its texture. If it feels gritty, then it's high in sand. If it feels slippery, it's high in silt. If it's sticky when wet, that's clay, which is good for pots and bricks but bad for drainage. Is it full of sand and rocks? This soil originated in a riverbed or glacial moraine and will be hard to dig. If your soil is sandy and full of rocks, it's probably well drained. In this case, consider spreading rain across the existing landscape rather than digging a basin.

Two simple tests will tell you a lot about what

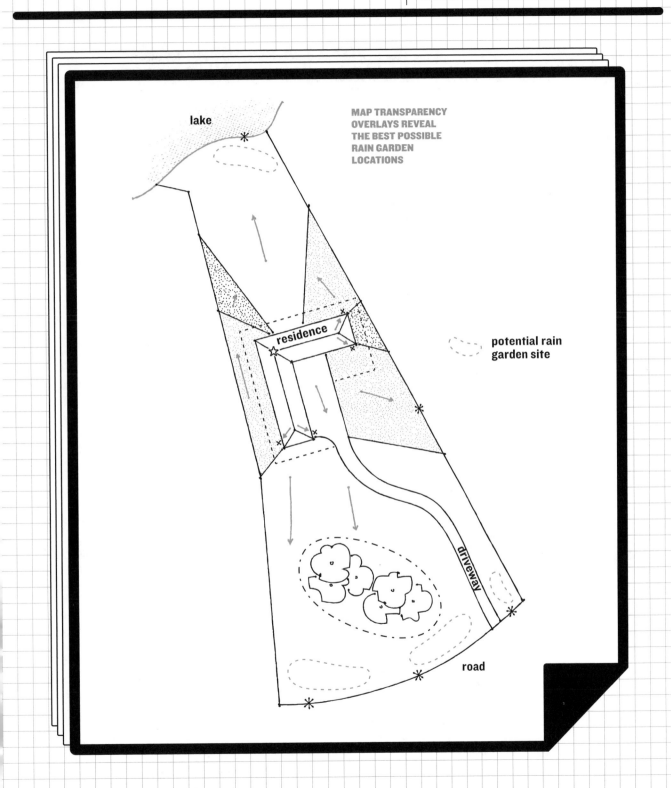

lake

MAP TRANSPARENCY
OVERLAYS REVEAL
THE BEST POSSIBLE
RAIN GARDEN
LOCATIONS

residence

potential rain
garden site

driveway

road

THE BEST TYPES OF RAIN GARDEN BASINS FOR VARIOUS SOIL TYPES

U.S. DEPARTMENT
OF AGRICULTURE SOIL
CLASSIFICATION CHART

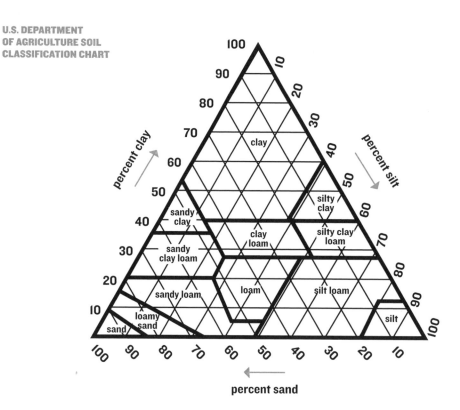

percent clay

percent silt

percent sand

SOIL TYPE	APPROXIMATE MINIMUM DRAINAGE RATE	RECOMMENDED TYPE OF RAIN GARDEN BASIN
Sand	8.25 inches/hour 21 cm/hour	Shallow basin with compost mixed into garden soil to improve water retention. Use drought-tolerant plants.
Sandy loam	1.00 inch/hour 2.5 cm/hour	Shallow basin with compost mixed into garden soil
Silt loam	0.25 inch/hour 0.6 cm/hour	Shallow to moderate-sized basin with compost mixed into garden soil
Clay loam	0.10 inch/hour 0.25 cm/hour	Large basin with amended soils. Consider plants with deep taproots to help break the clay and establish better drainage.
Clay	0.02 inch/hour 0.05 cm/hour	Not recommended. Choose a site with better drainage or design a pond.

PROJECT

TEST YOUR DRAINAGE RATE

Engineers use an infiltration test (sometimes called a percolation or perk test) to measure drainage at a given depth. A rain gardener needs to know how quickly a rain garden will drain when the soil underneath is saturated with water. Based on this drainage rate, you can check the table to figure out what type of rain garden basin best suits that site.

MATERIALS

SHOVEL
STAKE OR STICK
WATER
MARKER
TIMEKEEPING DEVICE
MEASURING TAPE OR RULER

1. Dig a 1-foot (30-cm) test hole in well-draining soil or a 2-foot (60-cm) hole in slowly draining soil. If you're doing the infiltration test when the soil is dry, fill the test hole with water, let it drain, then repeat three times.

2. Jab the stake or stick into the middle of the test hole.

3. Fill the hole with water.

4. Mark the filled water level on the stake with the marker.

5. Note the time.

6. Wait for the hole to drain completely, then note the time again.

7. Measure the distance from the mark on the stick to the bottom of the hole (in inches or cm). This is the water depth.

8. Calculate how long (in hours) the water took to drain. This is the drainage time.

9. Divide the water depth by the drainage time. The result is the drainage rate (inches /hour or cm/hour).

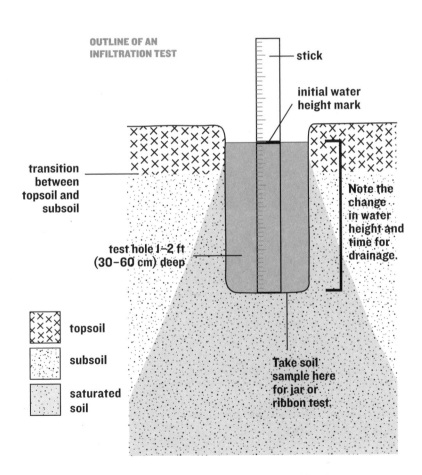

OUTLINE OF AN INFILTRATION TEST

stick

initial water height mark

transition between topsoil and subsoil

Note the change in water height and time for drainage.

test hole 1–2 ft (30–60 cm) deep

Take soil sample here for jar or ribbon test.

topsoil

subsoil

saturated soil

CASE STUDY

SYNCLINE RESIDENCE RAIN GARDENS Karla Dakin

PLACE: Boulder, Colorado, U.S.A.

DESIGNER: Karla Dakin, K. Dakin Design Inc.

INSTALLED: March 2009

RAIN GARDEN SIZE: 77 square feet (7 square meters) and 26 square feet (2.5 square meters)

ANNUAL RAINFALL: 19 inches (48 cm)

THIS HOUSE SITS at the base of the Boulder foothills, abutting open space at the bottom of the Rocky Mountains. Not only is the home collecting drainage from a huge area, it technically sits in a flood zone by a stream. Drainage was at the forefront of our consciousness in designing not only the landscape but also the architecture. There are two sump pumps in the basement, a living roof collecting and

Detail of the rain garden with the undulating stone and gestures of horsetail, Japanese blood grass, and blue iris.

filtering most of the water from the roof, and these two beautiful rain gardens in front of the house. Instead of trying to make the water disappear, we turned this problem into a fantastic design solution. At a minimum, 80 percent of the water that comes onto the site drains into the two rain gardens that delineate the front entry, creating unique sculptural forms.

What I really love about these rain gardens—in addition to the vital service they perform to slow water down and filter it before it leaves the site—is that they are made of material from the construction site. All around the house, inside and out, there are elegant blue stone patios. The left over stone was placed on its side to create undulating cobble. I worked with the stone mason to figure out the best way to install the stone, with both of us deciding to forego the substrate of sand and work directly on the dirt. The undulating forms remind me of the nearby topography of the foothills as they soften to the Great Plains east and beyond.

We left small holes in which we planted water-loving plants: horsetail (*Equisetum*), an ancient plant growing native in the nearby stream, Japanese blood grass (*Imperata cylindrica* 'Rubra'), and blue Siberian iris (*Iris sibirica*). The roaming nature of the Equisetum is curtailed by the stone. Low limestone walls flank the garden, tying into the the limestone veneer of the house. I look forward to seeing the plants grow in, bringing accents of color against the blue-gray backdrop.

Front entry of the residence showing rain gardens in the fore- and background.

kind of soil you have. The ribbon test shows how much clay is in the soil, and the jar test shows the relative amounts of sand, silt, clay, and organic matter. We recommend you do both tests to get a good feel for your soil, but feel free to skip the jar test if you're short on time. For each test, take soil from the bottom of a test hole.

To perform the ribbon test, dampen some soil and make a walnut-sized ball. Push the soil into a ribbon between your thumb and forefinger. If you can make a ribbon extend 2 inches (5 cm) or more without breaking, the soil is high in clay, which means it probably drains slowly. If the ribbon breaks when it is less than 1 inch (2.5 cm) long, the soil is very sandy. If the ribbon extends between 1 and 2 inches (2.5 and 5 cm), it's a mixture of sand, silt, and less than 20 percent clay known as loam. The soil chart on page 50 subdivides loam into sandy, silt, and clay; use your judgment to decide which best describes your soil, or continue on to the jar test.

The jar test is more precise but slightly more involved, requiring these seven steps:

1. Fill a glass jar (1 quart or 1 L works well) half full of soil.

2. Fill the jar with water and cap it.

3. Shake the jar vigorously until the soil turns into a uniform slurry. (Note: If your soil is very clayey, it may take hours or days for the clay lumps to dissolve. A few lumps are okay.)

4. As soon as you stop shaking, count 5 seconds and mark a line on the outside of the jar (a permanent marker or crayon works well) that shows the top of the coarse sand layer that should be visible at the bottom of the jar. This is the sand component.

5. Put the jar down and set a timer for 30 minutes. Come back and mark the level of settled sediment. This is the silt component.

6. Put the jar on a shelf and forget about it for several days or longer. When you come back, you will see a layer of clay particles above the silt layer and some black organic matter floating on top of the water. Mark the top of the clay layer.

7. **Estimate and record the relative percentages of sand, silt, and clay in the jar.**

Using the percentages of sand, silt, and clay you recorded from the jar test, find your soil type in the accompanying soil classification chart. Based on the results of the ribbon test and/or jar test, check the table to figure out what type of rain garden basin best suits your soil conditions. Note the approximate minimum drainage rate and type of rain garden basin listed in the table on the worksheet at the end of this chapter.

> ##
>
> ## LAST WORDS
>
> **RAIN GARDEN DESIGN** is a kind of looping spiral through exploration, goal setting, and site assessment. In the dreaming up phase, go back to your favorite nature trail or visit other rain, bird, or butterfly gardens for inspiration. Whatever sparks your imagination at the conservatory of flowers or the botanical garden could have a place in your rain garden. Make notes or snap photographs of these elements and add them into the mix. This design process continues even after the rain garden is in the ground. After a big storm, you may decide to make the basin larger. A plant sale might yield an irresistible new addition. Or you may develop new goals, like growing edible mushrooms or weaving sweetgrass baskets, and then add inoculated logs in the shade of a dogwood or plant sweetgrass along the rain garden fringe. Don't lose track of this looping process as you move into the more technical part of rain garden design. Keep in mind the rain garden of your dreams.

WORKSHEET

Rain Garden Information

AREA OF ROOF(S) _____

AREA OF HARDSCAPE (driveway, sidewalk, patio) _____

TOTAL CATCHMENT SURFACE AREA (roofs + hardscape) _____

ESTIMATED INFILTRATION AREA NEEDED (multiply total catchment surface area by 0.2) _____

RAIN GARDEN SITE 1	**RAIN GARDEN SITE 2**
SLOPE AT RAIN GARDEN SITE:	**SLOPE AT RAIN GARDEN SITE:**
RIBBON TEST RESULT: Ribbon length is inches (cm)	**RIBBON TEST RESULT:** Ribbon length is inches (cm)
JAR TEST RESULTS: Soil is % sand, % silt, % clay	**JAR TEST RESULTS:** Soil is % sand, % silt, % clay
SOIL TYPE:	**SOIL TYPE:**
DRAINAGE RATE:	**DRAINAGE RATE:**
APPROPRIATE BASIN TYPE:	**APPROPRIATE BASIN TYPE:**

CHAPTER 2

DESIGNING YOUR RAIN GARDEN

AFTER WORKING THROUGH

THE SITE ASSESSMENT EXERCISES IN THE last chapter, you should have identified several places where your rain garden(s) could go. Now let's walk through the technical details. In this chapter we explain:

▶ **How to calculate how much water will be flowing toward your rain garden**

▶ **How large of a rain garden is required to manage that amount of water**

▶ **How to get water from the roof, driveway, and other hardscape to the rain garden**

▶ **How to deal with excess rain so that it doesn't erode soil or cause a flood**

▶ **How to make sure your rain garden doesn't breed mosquitoes**

PREVIOUS SPREAD: This rain garden fits between the house and driveway.

We've geared our design process for the rule-of-thumb gardener, who relies on observation and tinkering. The Engineer boxes explain optional activities for the gardener who loves calculations. These calculations account for rain that splashes out of the gutter or infiltrates before it reaches the basin, so the estimates of basin size will be more precise. If you skip the Engineer sections, you'll need to keep a close eye on your rain garden for a few years to see if it's large enough to deal with the most torrential storms. If you notice water overflowing the basin every time it rains, you should expand it.

As you work through the exercises that follow, keep track of the results of your calculations on the information worksheet at the end of the chapter.

MEASURING YOUR CATCHMENT

It's important to figure out how much water is actually running off your property. The first step is to take a surface inventory by noting all the rooftops, driveways, and patios—all the impermeable surfaces where rain will fall and run off. The size of these surfaces is the drainage area.

Next determine what portion of these surfaces slope toward your planned rain garden site. If you're planning multiple rain gardens, calculate the total square footage of drainage areas for each. For example, half of the roof may slope toward the front yard, the other half toward the backyard. It's usually easiest to make two rain gardens, each receiving half the

AVOIDING MOSQUITO PROBLEMS

THE LAST THING you want in your garden is a malarial swamp. Mosquitoes breed in water and moist soil. And that's what you get in a rain garden, right? Not necessarily. The fastest a mosquito can develop from an egg into a biting, disease-spreading menace is four days. In a mosquito-proof rain garden, water drains before mosquitoes mature: at most, 48 hours after the most intense storms. If your rain garden drains in less than two days, it can't breed mosquitoes or contribute to the spread of malaria or West Nile virus.

CASE STUDY

A PRAIRIE RAIN GARDEN FOR THE BIRDS AND BUTTERFLIES

by Josh Simerman

PLACE: Fort Wayne, Indiana, U.S.A.

HOMEOWNERS: Kurt and Pam Simerman

DESIGNERS: Josh and Leara Simerman

INSTALLED: August 2009

CATCHMENT AREA: 850 square feet
(80 square meters)

RAIN GARDEN SIZE: 325 square feet
(30 square meters)

ANNUAL RAINFALL: 36 inches (90 cm)

INSPIRED BY natural gardeners like Masanobu Fukuoka and Sepp Holzer, we transformed our family's lawn into a complex yet self-sustaining web of rain garden basins and channels. We wanted to add natural beauty, variation, and texture near a screened porch; fix drainage problems including flooding near the house and standing water at the property line; and create habitat for frogs and crickets (to provide summertime evening music), birds, bees, butterflies, and hummingbirds. As the garden grows, it will also provide a natural play and learning space for young family members in future years.

The rain garden includes many wildlife features. A rock channel spills into a small holding pond, and overflow continues down the slope to a large evaporating pool. The holding pond houses mosquitofish and water plants. It creates a cool area in the garden and a watering hole for small animals. Rocks and boulders in the rain garden basin collect heat from the sun, speeding evaporation and creating a humid microclimate that attracts mosquito predators such as dragonflies. Large logs and bundles of branches cross the rock channel, providing small animal corridors and places for beneficial fungi to grow. Several bird houses above the garden encourage birds to populate the area, reducing insect pests and providing fertilizer. A bat box shelters bats, which also help to manage mosquitoes. Native bees nest in a nesting pole.

We broke ground in August 2009 and finished planting the following spring. We dug up the lawn in the rain garden area and then built up berms from the sod mixed with clay and soil from the channels. The grass and roots composted inside the berms over the winter and provided a very fertile planting area for

Winter rain garden plants continue to provide interest and beauty to the landscape.

CONTINUED

the spring planting. In August, once we had dug and shaped the berms, mounds, and channels, we placed river rock and planted a peach tree (*Prunus persica*), elderberry tree (*Sambucus*), and large shrubs such as a French lilac bush (*Syringa vulgaris*) for fragrance and a butterfly bush (*Buddleia americana*). We scattered seed for other annuals and perennials. In early spring, we touched up the river rock, adding more rocks to the channel where rock had settled into the soil, mulched the entire area, and planted spring-dormant plants and seeds, including aromatic aster (*Symphyotrichum oblongifolium*), queen of the prairie (*Filipendula rubra*), and cardinal flower (*Lobelia cardinalis*).

Digging out the initial garden area, bringing in several tons of river rocks to line the channels, and spreading several yards of cedar mulch was by far the greatest challenge, in terms of labor and money. The soil at the site is heavy dense clay left behind by receding glaciers, which is rich in minerals but also very difficult to dig into. We often had to wait several days for a bit of rain to soften it up before we could continue to dig. Because of the challenging digging conditions, we did not dig very deeply into the ground.

Over time, the plants improved drainage. So far, the blackberry bushes (*Rubus*) have grown the fastest, and the berries attract small animals to the garden. The butterfly bushes got rather large in their first

year. Both plants can be invasive in some areas, but we've found them easy to cut back. Sedges have grown well in the wet zone, doubling in size during a single summer. The sedges provide nesting materials for birds and an interesting autumn texture that remains after the other plants have died back. Prairie natives and coneflowers have done well in damp areas.

Even in the first year, that area of the landscape has changed dramatically. On spring mornings, dozens of mushrooms pop up from the mulch. The air seems fresher and cooler on hot days in the summer, and the moisture definitely attracts many more crickets, toads, and small frogs. Bright flowers entice neighbors over to chat. In autumn, the garden traps drifting leaves, which break down so

quickly there's no need to rake them. The greenery of the rain garden plants often lasts much longer than the rest of the yard, even into mid-December, when the hard freezes turn everything a deep reddish brown.

We've made an effort to manage and harvest all of the rainwater that falls onto the property so that as little water is directed to sewers and off-property drainage as possible. Since the original installation, we added two (much smaller) rain garden beds in open areas. During some heavy rains we examined how water flowed from downspouts into the surrounding landscape and out onto the grass. From these observations, it was easy to see how water flowed through the yard, so we dug out additional shapes to catch the water on its way to the street.

The rain garden encourages wildlife by providing microhabitats and produces edibles throughout the season.

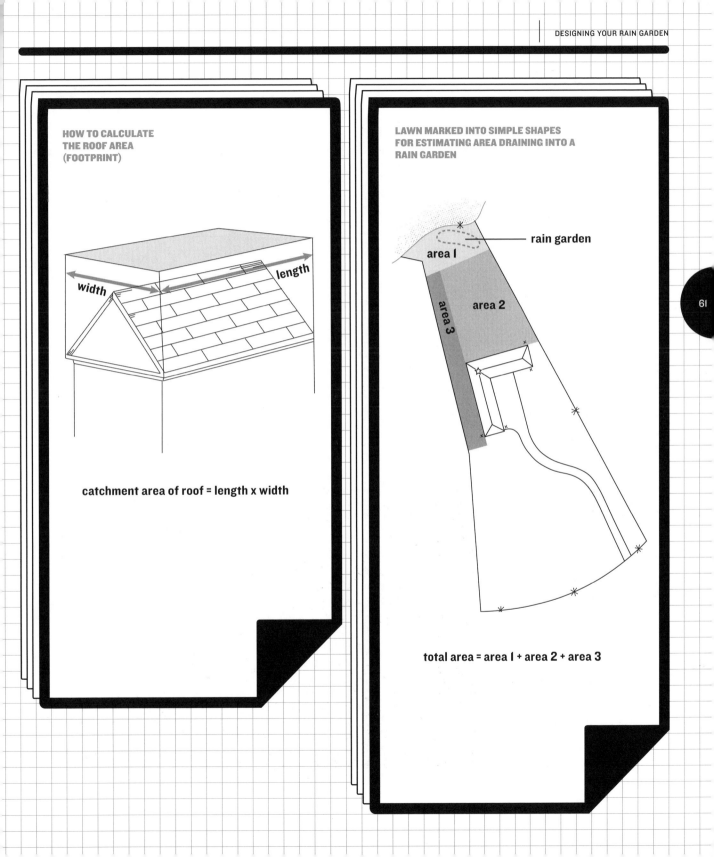

**HOW TO CALCULATE
THE ROOF AREA
(FOOTPRINT)**

width

length

catchment area of roof = length x width

**LAWN MARKED INTO SIMPLE SHAPES
FOR ESTIMATING AREA DRAINING INTO A
RAIN GARDEN**

rain garden

area 1

area 2

area 3

total area = area 1 + area 2 + area 3

61

An inlet entering on a splash guard of stone disguised by ferns and grasses.

roof runoff. In this case, consider only half of the roof area as a catchment surface for each rain garden.

Should you consider lawn or landscaping as part of your catchment area? It depends. Lawns and landscaped areas often have compacted soils that infiltrate some runoff but let about two-thirds of the water run off. If roof runoff will have to cross 30 feet (9 m) or more of lawn to reach the rain garden, you should consider the lawn an additional catchment area. If landscape or lawn is small compared to the total catchment area, you may ignore it in your calculations.

When calculating the area of catchment surfaces, measure the footprint—don't worry about peaks, valleys, or slope. Using an aerial image of your property, such as one from Google Earth, helps you to see and measure the flat footprint of each area. If you don't have an accurate site plan or access to Google Earth, you can measure the roof footprint by measuring the length and width of your house with a tape measure.

Although it may seem difficult to measure the area of lawn draining into a rain garden, remember that you don't need to be exact. Mark simple lines across the lawn with chalk or on your base map to break up the landscape into rectangles, triangles, and circles. Here are some helpful equations for calculating surface area:

▶ **Area of a square or rectangle = length × width**

▶ **Area of a right triangle = 0.5 × base × height**

▶ **Area of a circle = π × radius² = 3.14 × radius × radius (note: the radius is half the diameter)**

To calculate the total surface area draining into a rain garden, first calculate the area of each catchment surface, then add all the areas together:

▶ **Area 1 (roof) = length × width × percent directed to rain garden**

▶ **Area 2 (driveway) = length × width × percent directed to rain garden**

▶ **Total drainage area to rain garden = Area 1 + Area 2**

The percent of rain directed to the rain garden is the amount that flows down a particular downspout. If three downspouts drain a roof surface, in theory, 33% of rain falling on that roof would be directed to the rain garden. The percentages should be in decimal format; for example, for 25% multiply by 0.25 and for 100% multiply by 1.00.

CALCULATING THE FRACTION OF RAIN THAT RUNS OFF

THE SIZE OF THE CATCHMENT AREA is only part of the equation. Not all rain that falls on a surface will run off—some will splash out of the gutter, evaporate, or sink into the soil. Hydrologists call the fraction of rain that runs off the runoff coefficient. As you might guess, pavement has a higher runoff coefficient than lawn. Deep-rooted, multilayered landscapes such as forests have low runoff coefficients, because most rain gets caught on leaves and sinks slowly into the soil. The table provides runoff coefficients for some common surface materials.

EXAMPLES OF RUNOFF COEFFICIENTS

SURFACE	TYPICAL RANGE	RECOMMENDED VALUE
Concrete	0.80–0.95	0.90
Brick	0.70–0.85	0.80
Roofs	0.75–0.90	0.85
Paving stones	0.10–0.70	0.40
Grass pavers/turf blocks	0.15–0.60	0.35
Lawns and grass on sandy soil	0.05–0.20	0.12 for a gentle (5%) slope, with values increasing as slope increases
Lawns and grass on heavy soil	0.13–0.35	0.22 for a gentle (5%) slope, with values increasing as slope increases
Landscaped beds	0.15–0.30	0.20
Crushed aggregate	0.15–0.30	0.20

Adapted from LEED-NC version 2.1

Applying runoff coefficients in your calculations will yield a more accurate estimate of how much water will flow into the rain garden. This is known as the effective runoff area. If you choose not to use a runoff coefficient in your calculations, you'll end up with a slightly larger rain garden than you need.

For example, here's how to calculate the effective runoff area for a house that has an asphalt shingled roof and concrete driveway. Let's assume the roof is 25 feet (7.5 m) by 40 feet (12 m) and the driveway is 10 feet (3 m) by 16 feet (4.8 m). Use the runoff coefficients for roofs and concrete from the table.

I. CALCULATE THE EFFECTIVE RUNOFF AREA OF THE ROOF AND THE DRIVEWAY.

▶ **Effective runoff area = length × width × percent directed to rain garden × runoff coefficient**

▶ **Area I (roof) = 25 feet (7.5 m) × 40 feet (12 m) × 0.50 × 0.85 = 425 square feet (38.3 square meters)**

▶ **Area 2 (driveway) = 10 feet** (3 m) × 16 feet (4.8 m) × 1.00 × 0.90 = 144 square feet (13 square meters)

2. ADD THE TWO AREAS TOGETHER TO GET THE TOTAL EFFECTIVE RUNOFF AREA.

▶ **Area I + Area 2 = 425 square feet (38.3 square meters) + 144 square feet (13 square meters) = 569 square feet (51.3 square meters)**

TYPICAL RAIN GARDEN DEPTH BASED ON SOIL TYPE AND TOTAL CATCHMENT AREA

SOIL TYPE	SIZE OF RAIN GARDEN	RECOMMENDED EXCAVATION DEPTH
Sandy loam	10% of catchment area	6 to 12 inches (15 to 30 cm)
Silty loam	20% of catchment area	12 to 18 inches (30 to 45 cm)
Clayey loam	30% of catchment area	18 to 24 inches (45 to 60 cm)

ESTIMATING THE SIZE OF YOUR RAIN GARDEN

If you're an Engineer, you'll be able to calculate rain garden size fairly precisely, using measurements you've already compiled. If you're a tinkerer and not in the mood for math, we offer easy rules of thumb that work well for moderate rainfall on loamy soils. If you live in a very wet or dry climate or if your soil drains quickly or very slowly, we strongly encourage you to double check these guidelines against a local rain garden guide. The world won't end if your rain garden is too large or too small, but following local practices will give your rain garden the greatest impact and chance of success.

Stormwater engineers have come up with easy rules of thumb for sizing rain gardens. These rules give reasonable estimates of the runoff volume and pooling depth, two factors that you need to know in order to design a rain garden

basin. Runoff volume is how much rain will run off the catchment surface during a large storm, for example, 2 inches (5 cm) of rain in 24 hours. Pooling depth is related to how quickly water drains in the basin. Let's start by assuming you'll try to infiltrate all the runoff from your catchment area.

If you've been following along with previous exercises, you have calculated the size of the catchment area (the hard surfaces that will drain into the rain garden) and determined your drainage rate.

As you found with the ribbon, jar, or infiltration test, soil type determines drainage rate. Water sinks quickly into sand and seeps slowly into clayey soil. Most soils contain a mix of sand, silt, and clay known as loam. In typical loamy soils, which are 40 percent sand, 40 percent silt, and 20 percent clay, you'll need 1 square foot (0.1 square meter) of rain garden for every 5 square feet (0.5 square meter) of catchment area. In other words, the surface area of the rain garden will be 20 percent of the catchment area. The sandier the soil, the smaller the rain garden can be—down to about 10 percent of the drainage area. Clay soils, on the other hand, will require a larger rain garden, sometimes even larger than 30 percent of the drainage area. If you have heavy clay soil and don't want such a large rain garden, you can amend your soil with compost and add deep-rooted plants, which will eventually increase the drainage rate. If you have very sandy soil and want to retain water longer, you can loosen the soil to about 6 inches (15 cm) and mix in 3 inches (7.5 cm) of woody compost.

The depth of a rain garden basin also depends on the soil type. The table lists a range of excavation depths for each soil type. In slow-draining sites, many rain gardeners dig an 18- to 24-inch (45- to 60-cm) basin, then fill the basin with amended soils to 6 inches (15 cm) below the outlet, so that the basin has a 6-inch (15-cm) pooling depth and doesn't look like a deep pit.

These are generally conservative guidelines. Following them will usually create a basin large enough to capture the rain from most storms. Even so, we recommended you keep a close eye on your rain garden for the first season to see how often it fills up and overflows and how long it takes to drain after a storm. If you have a small property, you may have to scale back on the ambition of harvesting all the runoff from your catchment area. But a rain garden of any size will reduce stormwater pollution. If your rain garden is designed to harvest just a portion of your runoff, it is vital to read the section "Getting Water into and out of the Rain Garden" at the end of this chapter to ensure excess water is directed to a safe and appropriate location.

ESTIMATING HOW MUCH WATER WILL FLOW TO YOUR RAIN GARDEN

When rain falls on your house, water washes down the roof toward the gutters. Some gets soaked up by the shingles or splashes out of the gutters, and the rest runs down the downspout and sinks in the soil, runs off onto the landscape, or surges downhill to a storm drain. A rain garden basin collects rain during a storm, and well-draining soils help runoff sink into the soil between storms.

Rain begins infiltrating as soon as it reaches the rain garden basin. If rain falls gently, the basin drains faster than the water flows in, and no pooling should occur. In a downpour, rain will rush into the basin faster than it can infiltrate into the soil and your rain garden will begin to fill. Once rain stops falling, the level of water in the basin drops—quickly in sandy soils, more slowly in clayey or compacted soils. Ideally, a rain garden fills up during intense storms, but doesn't overflow, and then drains completely within 24 hours after a storm.

USING A RAIN GAUGE

A RAIN GAUGE is an easy way to observe and collect rainfall data for your specific site. You don't need to do these measurements before you design your rain garden, but they will help you evaluate how effective your rain garden is at capturing the local precipitation. You can buy a rain gauge at many garden supply stores. Place the rain gauge in an open area free of trees or structures to ensure nothing blocks the rainfall. After every rain, record the level of the rain in the gauge, then dump it out so you're prepared for the next storm. Also note the date and approximately how long the storm lasted. You'll need to record rainfall for at least an entire rainy season if you want to use the data in your rain garden calculations—the longer the record, the better. At the end of the rainy season, look at how much rain fell during the three largest storms. Pick one of these values to use in calculating the volume of your rain garden basin, or do separate calculations with each of the highest rainfall values and see how much the volume of runoff to the rain garden varies.

HOW TO READ A RAINFALL CHART

TO DISCOVER HOW MUCH RUNOFF you're dealing with—and how large to make your rain garden—you need to know how much rain falls during the largest storm event you want to capture. The Engineer will use rainfall measurements to design the basin, while the rule-of-thumb gardener can observe when the rain garden overflows and then see how much rain fell that day. The most important measurements for rain garden design are how much rain falls in a large storm and how long storms typically last.

Meteorologists collect a dizzying array of rainfall data—from 15-minute peak intensities to running averages of daily, weekly, and monthly rainfall. In most cases, use 24-hour average rainfall data. If storms in your area are flashy, with most rain falling in intense cloudbursts, you will want to know how much rain falls in a 15- or 30-minute downpour. If you live in a place where it rains more often than it doesn't, figure out how long the longest storms last and how many clear days you can expect between storms to see if your rain garden will have a chance to drain between storms.

To obtain rainfall charts, contact your local weather bureau or agricultural extension agent and ask them to provide the data in chart form. The rainfall charts show rainfall intensity for different periods of time. The vertical axis shows how much rain fell, and the horizontal axis shows time intervals. In the 24-hour rainfall chart, we see that the storm on day 7 produced the most rainfall in 24 hours, about 5 inches (12.5 cm). Based on this chart, the rainfall intensity would be 5 inches over 24 hours or 0.2 inches (0.5 cm) per hour. In the 1-hour chart, we see that the most rain fell during the 22:00 period, about 0.5 inches (1.3 cm) per hour. In the 15-minute rainfall chart, we see that the most rain fell during the 21:45 interval, about 0.15 inches (0.38 cm) over 15 minutes or 0.60 inches (1.5 cm) per hour. Therefore, the 15-minute rainfall totals result in the greatest estimated rainfall intensity. You can also use 1-hour or 15-minute rainfall data to estimate the duration of a typical storm. In the 1-hour and 15-minute rainfall charts, we see a 4-hour downpour from 12:00 to 16:00 and a 1-hour cloudburst beginning at 21:30.

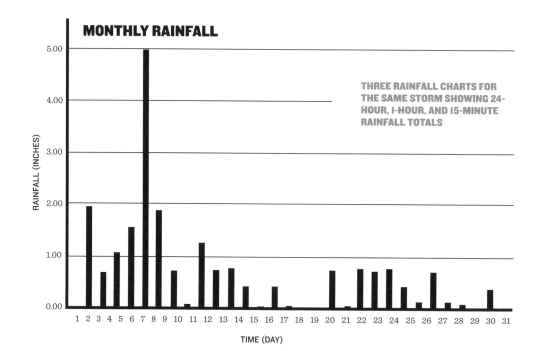

MONTHLY RAINFALL

THREE RAINFALL CHARTS FOR THE SAME STORM SHOWING 24-HOUR, 1-HOUR, AND 15-MINUTE RAINFALL TOTALS

RAINFALL (INCHES)

TIME (DAY)

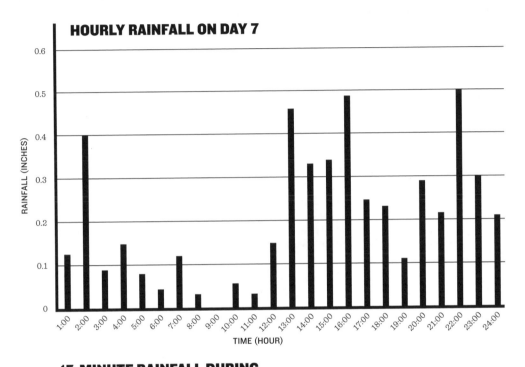

HOURLY RAINFALL ON DAY 7

RAINFALL (INCHES)

0.6
0.5
0.4
0.3
0.2
0.1
0

TIME (HOUR)

1:00 2:00 3:00 4:00 5:00 6:00 7:00 8:00 9:00 10:00 11:00 12:00 13:00 14:00 15:00 16:00 17:00 18:00 19:00 20:00 21:00 22:00 23:00 24:00

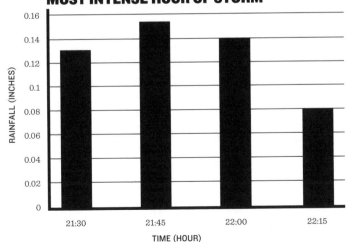

15-MINUTE RAINFALL DURING MOST INTENSE HOUR OF STORM

RAINFALL (INCHES)

0.16
0.14
0.12
0.1
0.08
0.06
0.04
0.02
0

21:30   21:45   22:00   22:15

TIME (HOUR)

ENGINEERS / You will need to look at a few more days of data to zero in on an average duration to use in the Engineer calculations. If you can't track down hourly or 15-minute data for your area, pick a very rainy day and use the 24-hour rainfall depth and 24 hours for the duration of heavy rainfall.

CALCULATING RUNOFF VOLUME AND RAIN GARDEN DIMENSIONS

HOW MUCH WATER runs off of your landscape—the runoff volume—depends on catchment area, a runoff coefficient, and the rainfall intensity. Rainfall intensity is the amount of rain that falls over a given period of time (for example, 2 inches per hour or 15 cm per 24 hours) during the most intense phase of a storm. Stormwater engineers typically use the most intense rainfall of a heavy storm (the peak intensity) in calculations, and use this design storm intensity to size basins for a peak "1-year" or "25-year" storm. The larger this design storm, the larger the basin will be. Choosing which value to use is an inexact art, as it depends on the time it takes a raindrop to travel from roof to rain garden, among other factors. For home rain gardens, there is no perfect storm and no real need to be so precise. You can choose a rainfall intensity value three ways: tracking rainfall with a rain gauge, looking at rainfall charts that show daily or hourly rainfall for your region, or asking your local public works department what design storm rate they use. We recommend considering a range of values reported for your region to see how they affect your calculated runoff volume.

DETERMINING THE VOLUME

To determine dimensions for your rain garden, first you need to calculate the runoff volume. Let's assume you have 1000 square feet (93 square meters) of catchment area directed to your rain garden and your local public works department tells you they use a rainfall intensity of 1 inch (25 mm) per hour as average design storm intensity. From experience, you know that heavy rainfall in your area usually lasts around 1 hour. Use the following equation to convert the catchment area to a volume of water draining off during a 1-hour storm:

**Runoff volume (cubic feet) = total drainage area (square feet) ×
[rainfall intensity (inches/hour) × (1 foot/12 inches)] × duration of heavy rainfall (hours)**

$$\text{Runoff volume} = 1000 \text{ square feet} \times \left[\frac{1 \text{ inch}}{1 \text{ hour}} \times \frac{1 \text{ foot}}{12 \text{ inches}} \right] \times 1 \text{ hour} = 83.3 \text{ cubic feet}$$

In this calculation, first do the multiplication within the brackets and then multiply that value by the square footage.

For metric calculations, use this equation:

**Runoff volume (cubic meters) = total drainage area (square meters) ×
[rainfall intensity (mm/hour) × (1 m/1000 mm)] × duration of heavy rainfall (hours)**

$$\text{Runoff volume} = 93 \text{ square meters} \times \left[\frac{25 \text{mm}}{1 \text{ hour}} \times \frac{1 \text{m}}{1000 \text{mm}} \right] \times 1 \text{ hour} = 2.3 \text{ cubic meters}$$

So the volume of the rain garden needs to be 83.3 cubic feet (2.3 cubic meters) in order to hold the runoff from a storm that delivers a peak rainfall of 1 inch (25 mm) of rain per hour for 1 hour.

If you want to know your local average rainfall intensity, you'll need to do a bit of research. Rainfall data can be a bit tricky to track down and interpret, but your search will lead you to local experts who can answer questions that will come up later in the design process. Stormwater engineers are interested in intensity (how hard rain falls), duration (how long each storm lasts), and frequency (how much time elapses between storms). These factors all affect how quickly a rain garden will fill up and whether all rain will infiltrate between storms. If you use the local design storm rainfall intensity value in your rain garden design, you'll know your rain garden can handle the most intense storms and rarely overflow.

If you have trouble tracking down local rainfall intensity information, the U.S. Geological Survey, the National Oceanic and Atmospheric Administration, and Australia's Bureau of Meteorology produce graphs of daily average rainfall. These charts can be used to choose the average daily rainfall depth to capture in your rain garden. As a last resort, if you can't track down local data, use a rainfall intensity of I inch (2.5 cm) per hour, and you'll capture rain from most large storms.

DETERMINING THE DEPTH

The next step is to determine the ideal depth for your rain garden, then calculate its surface area using this depth and the calculated runoff volume. In loamy soils, a rain garden is between 6 and 9 inches (15 to 22.5 cm) deep. In very sandy soils, the rain garden may be only 3 inches (7.5 cm) deep. In slow-draining soils, you can dig a basin up to 24 inches (60 cm) deep; the extra depth will allow you to capture lots of runoff. But in slow-draining soils, 24 inches (60 cm) of water could take a week to infiltrate, creating a breeding ground for mosquitoes. The solution is to fill the basin almost all the way to the surface with a fast-draining soil mix, leaving a 6- to 9-inch (15- to 22.5-cm) depression. The distance from the soil surface to the bottom of the rain garden rain garden outlet is known as the pooling depth because it determines how deeply water pools in the basin.

How do you determine the best pooling depth? Soil texture is the key. The ribbon test, jar test, or infiltration test allowed you to estimate the soil's drainage rate. Because a rain garden should be able to infiltrate all pooling water within 24 hours, convert the drainage rate from inches (cm) per hour into inches (cm) per day. For example, let's assume that during an infiltration test, the water in the hole dropped I inch (2.5 cm) in 6 hours.

Thus, if your rain garden is 4 inches (10 cm) deep from the bottom of the basin floor to the bottom of the outlet, it should be able to infiltrate all of the water that falls into it within a 24-hour period. If you make the rain garden deeper than that, you may collect more water than it is able to drain within 24 hours.

You control pooling depth by changing the height of the outlet. The outlet diverts overflowing water out of the basin along a path you choose, either through a surface drain or a more complicated under drain. The vertical distance from the bottom of the basin to the bottom of the outlet will determine how deep the water pools in the basin. See the end of this chapter for more on overflow design.

Maximum pooling depth (inches or cm) = soil drainage rate (inches/hour or cm/hour) x 24 hours/day

$$\text{Maximum pooling depth} = \left[\frac{1 \text{ inch}}{6 \text{ hours}} \quad \text{X} \quad \frac{24 \text{ hours}}{1 \text{ day}} \right] = 4 \text{ inches/day}$$

$$\text{Maximum pooling depth} = \left[\frac{2.5 \text{cm}}{6 \text{ hours}} \quad \text{X} \quad \frac{24 \text{ hours}}{1 \text{ day}} \right] = 10 \text{ cm/day}$$

CONTINUED

◢ **CONTINUED**

DETERMINING THE SURFACE AREA

Because volume is the product of area and depth, it's easy to calculate your rain garden's surface area once you've determined its ideal volume and depth. The pooling depth you calculated in the previous equation was in inches or centimeters, so here you need to convert to feet or meters.

$$\text{Rain garden area (square feet)} = \frac{\text{rain garden storage volume (cubic feet)}}{\left[\text{rain garden pooling depth (inches)} \times \left(\frac{1\text{ foot}}{12\text{ inches}}\right)\right]}$$

$$\text{Rain garden area} = \frac{83.3\text{ cubic feet}}{\left[4\text{ inches} \times \frac{1\text{ foot}}{12\text{ inches}}\right]} = 250\text{ square feet}$$

$$\text{Rain garden area (square meters)} = \frac{\text{rain garden storage volume (cubic meters)}}{\left[\text{rain garden pooling depth (cm)} \times \left(\frac{1\text{ m}}{100\text{ cm}}\right)\right]}$$

$$\text{Rain garden area} = \frac{2.3\text{ cubic meters}}{\left[10\text{ cm} \times \frac{1\text{ m}}{100\text{ cm}}\right]} = 23\text{ square meters}$$

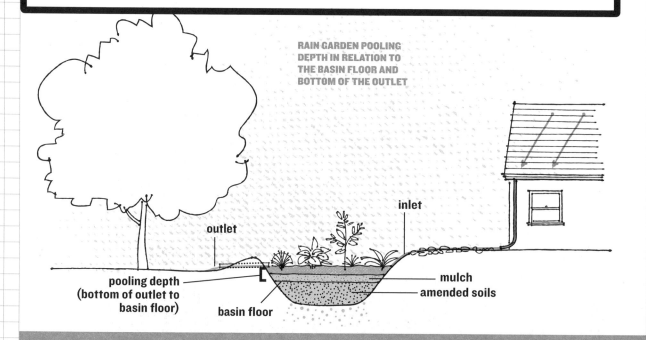

RAIN GARDEN POOLING DEPTH IN RELATION TO THE BASIN FLOOR AND BOTTOM OF THE OUTLET

outlet

inlet

pooling depth (bottom of outlet to basin floor)

basin floor

mulch

amended soils

This rain garden directs overflow away from the house through a rock-lined diversion swale to the street.

DETERMINING LENGTH AND WIDTH OF YOUR RAIN GARDEN

You now know the surface area and depth of your rain garden. In addition, you may already know the length or width of your garden, for example, if it needs to fit between the house and a sidewalk. Look at your landscape plan or go outside and note where existing features limit length or width. Divide the rain garden area by this limiting dimension to find the remaining dimension.

$$\text{Length} = \frac{\text{area}}{\text{limiting width}}$$

$$\text{Width} = \frac{\text{area}}{\text{limiting length}}$$

CHOOSING A SHAPE FOR YOUR RAIN GARDEN

Your rain garden can be any shape you choose, so get creative. Rounded or amoeba-like shapes create lots of edge, which is important for the myriad plants that thrive in boundary zones. If it's difficult to calculate the area of your created shape, divide it into approximate rectangles or circles, figure out the area of those shapes, and then add them together.

If you're dealing with a site that slopes between 5 and 20 percent or if you want to use your rain garden to irrigate a row of fruit trees or large shrubs, you can build a contour swale, which is a long, skinny rain garden. Contour swales follow contour lines, and their berms are perfectly level. (A contour line is an imaginary line used to indicate elevation across the landscape where all points falling on the line are at the same elevation.)

Contour swale basins are typically 18 to 24 inches (45 to 60 cm) wide, so use a width in this range to calculate how long to make the contour swales in order to infiltrate all of your runoff. Swales on slopes work well in series, so consider designing a cascade of swales that guides water from the top to the bottom of your landscape.

GETTING RAIN TO THE RAIN GARDEN

How you get rain to a rain garden depends on the site and your inclination. You can simply connect a downspout to a buried corrugated pipe or you can use rain chains, aboveground channels, and rock cascades to guide rain across the surface of your landscape. Here we describe how to move water across the landscape into the rain garden and how to best deal with overflow at times when the amount of rainfall is greater than your rain garden can handle.

TRACKING WATER FLOW

On a rainy day, watch the rain that's flowing from your downspout—or turn on a hose near the downspout—and watch where the water goes. Does it flow toward the rain garden site? If not, plan to redirect the water there with a pipe or channel from the downspout to the rain garden. If your driveway drains most water to the street (rather than to the side of the driveway), you'll need to install a drain across the driveway or remove a strip of concrete along the center and build a rain garden there. Consult an expert for assistance.

If water pools or barely flows toward your rain garden site, the ground may slope less than 2 percent. The pipe or diversion swale will still need to slope 2 percent in order for the water to flow, so on flat or uphill slopes, the pipe can end up being very deep. This makes for a pit-like rain garden, which is terrible for outdoor cocktail parties. On flat sites, put the rain garden basin close to the downspout, but no closer than 10 feet (3 m) from the foundation. You can also cut your downspout slightly above the ground and install a kicker, a lateral pipe that directs the water away from the house at a 2 percent downward slope (just don't let it be a tripping hazard).

MOVING WATER FROM ROOF TO LAND

Most homes with sloping roofs have gutters and downspouts, while flat roofs use long spouts called canales, scuppers, or drainage channels to direct rain away from building walls. If your house has no gutters or canales and you don't want to add them, the easiest strategy for moving water is to use a diversion swale.

If your downspouts are in working order, there's no reason to replace them. But if you're reroofing or need to move rainwater across a cement walkway, you might consider canales or rain chains to make the flow of rain off your roof more visually dramatic. Rain chains, arcing channels, and free-standing sculptures highlight water's path from roof to ground and add a dynamic visual—and sometimes musical—element to the landscape. Depending on your taste, choose bright copper cups to guide rain from a gutter to a flat stone, or let it run down an iron chain into a wooden rain barrel. If you're a sculptor, you could make ceramic rain chimes like Marina Wynton's or a free-standing sculpture to catch rain cascading out of a canale. Metal channels can arc over a sidewalk next to a building, eliminating the need to cut concrete.

This rain garden features a rock-lined diversion swale that moves water through the rain garden. This photo was taken during a 6-inch (15-cm) rainstorm that sent an estimated 4500 gallons (17,000 L) of water off the roof and into the rain garden without overflowing.

These rain chimes at Marina Wynton's home create melodies when it rains.

The garden also features a metal sculpture that splashes water from the roof canale into this living wall panel and the rain garden below.

VARIOUS WAYS TO CONNECT YOUR ROOF TO THE RAIN GARDEN

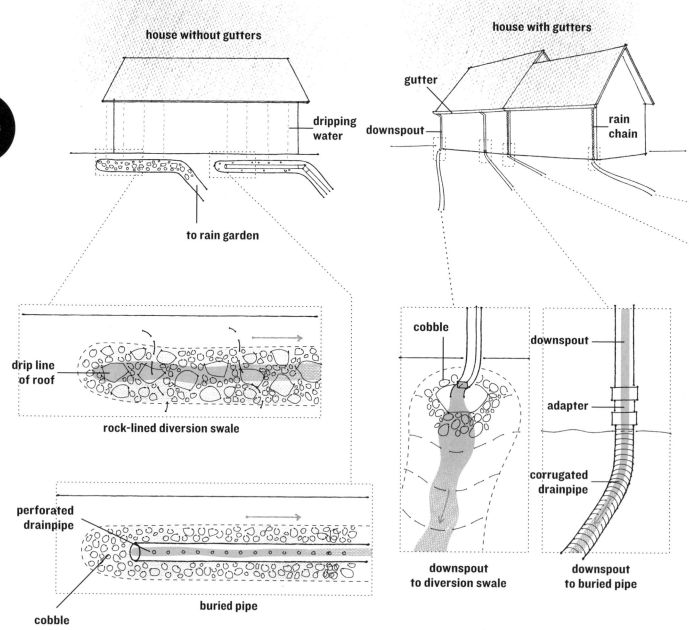

house without gutters

dripping water

to rain garden

house with gutters

gutter

downspout

rain chain

drip line of roof

rock-lined diversion swale

cobble

downspout

adapter

corrugated drainpipe

perforated drainpipe

buried pipe

cobble

downspout to diversion swale

downspout to buried pipe

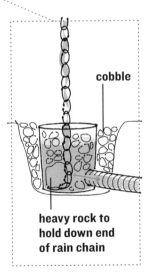

rain chain
to diversion swale

rain chain
to buried pipe

GETTING WATER TO THE RAIN GARDEN

Now that you've determined the size and location of your rain garden and guided rain from the roof to the ground, you need to move the water into the rain garden basin. For this, gravity is your greatest ally. Although you could install a sump and pump to get water from lower areas of a property to a higher rain garden, pumped systems are complex, expensive, and energy guzzling—and they inevitably break down. Gravity doesn't break down. The wisest choice is the path of least resistance: downhill.

There are five basic ways to move water to the rain garden, using sheet flow, a vegetated diversion swale, a rock-lined diversion swale, runnels (flumes), or buried pipe. Choose one of these methods or combine several, depending on your preferences and site conditions. Note that if you use sheet flow or diversion swales, some water will infiltrate before reaching the rain garden. For example, if your rain garden is far from the house and you use a diversion swale for the whole distance, water may only reach the rain garden after long, heavy rains.

For water to flow by gravity, the pipe or channel it flows through must slope at least 2 percent. Because flowing water causes erosion, keep channel slopes below 8 percent and take steps to minimize erosion wherever water flows over bare soil. On gentle slopes (1 to 5 percent), plants will slow water as sheet flow or in a diversion swale and keep the soil from washing away. In a rock-lined channel, the stones create a barrier that protects the soil from the erosive water rushing over it. If you can, choose a path for your swale that slopes gently. If your swale traverses ground that slopes more steeply, build rock, brick, or rubble steps.

Sheet flow describes water that spreads out over a large area, then moves as a thin film across the surface of the land without collecting into rills or channels. If your rain garden is relatively close to your drainage surface or if

77

the ground is graded so water flows to your rain garden site, you may not need to build any diversion structure. To check, lay a hose at the end of your downspout and turn it on. If most of the water flows to your rain garden site, all is well. If you notice eroding soil, add plants or stones between the drainage surface and the rain garden to slow the flow.

A diversion swale is a shallow, gently sloping channel that moves water across the landscape. A vegetated diversion swale is 18 to 24 inches (45 to 60 cm) wide and can be hundreds of feet long, and it is designed to move water instead of infiltrate it. (Some water does infiltrate as it flows down the diversion swale if the soil beneath is unsaturated.) A vegetated diversion swale uses grass or other dense, low-growing vegetation to keep flowing water from eroding the soil it flows over.

A rock-lined diversion swale uses cobble stone or round gravel to reduce the erosive power of flowing water. A rock channel usually needs less maintenance than a vegetated swale because there are no plants to care for, although you will need to weed regularly to keep water flowing freely. You can also guide water in runnels made of stone, cement, brick, or rot-resistant wood such as cypress or redwood.

If you want an entirely invisible waterway, use a buried pipe. When collecting roof runoff, a buried pipe can connect directly to a downspout and enter into the rain garden somewhere near the upper surface. When collecting from ground-level catchment surfaces like driveways or lawns, use an area drain to direct runoff to a buried pipe. The pipe should be about 4 inches (10 cm) in diameter; either black corrugated pipe (flexible) or rigid PVC drainpipe will work.

Why choose a buried pipe over a diversion swale? Slope is part of the answer. A pipe can move water down steep—or even vertical—slopes. Unlined swales must be laid out along a path that slopes between 2 and 4 percent, whereas swales lined with cobble and runnels

This corner rain garden catches sheet flows of runoff before the water can cascade over the sidewalk and into the street.

made of wood, cement, or metal can slope up to 8 percent. Build mini-cascades (also known as check dams) between sections of swale if you need to move water down steeper grades. If you have a path, patio, or other hardscape between your roof and rain garden, it's often easier to let water flow across the surface—or direct it through runnels—than to dig underneath.

GETTING WATER INTO AND OUT OF THE RAIN GARDEN

Flowing water will erode soil and mulch and eventually wash plants away. Inlets, where water enters the rain garden, and outlets, where it exits the rain garden, concentrate the flow of water and must be protected from erosion. Eroding soil and plant matter can clog inlet and outlet pipes, causing flooding and decreasing a rain garden's capacity. Therefore, careful design of inlets and outlets is key to a low-maintenance rain garden.

Think about how you will protect the soil around the inlet. Options include surrounding it with cobble, drain rock, or brick; placing a large flat stone beneath the inlet; or planting a dense groundcover. If you're using a pipe to move water into the rain garden, you can install a T fitting at the end to disperse the flow. You also may want to conceal the pipe with stone or plants. It's a good idea to plant sedges and rushes—or grasses in dry climates—with fibrous roots in front of the inlet to provide additional erosion control.

Outlets allow water out of your rain garden should it become so full it needs to spill over. No matter how carefully you have calculated your basin volume, one day a storm will blow in that exceeds all expectations. The smaller the rain garden, the more often your rain garden will overflow. It's important to give overflowing water a safe place to go. If you don't, gravity will choose the path for you, and water may end up where you least want it—like in your neighbor's yard or basement.

First, decide what kind of overflow to build:

a surface drain or an under drain. Surface drains are simpler to construct and work in most circumstances. But if your soils drain less than 0.1 inch/hour (0.25 cm/hour), an under drain may be necessary.

Surface drains

Surface drain options include vegetated or rock-lined diversion swales, runnels, and shallow buried pipe. On shallow slopes, sheet flow can work as well. Like the inlet, the outlet needs stone, brick, and tough vegetation around it to protect against erosion.

Where will water go when it overflows? Routing it to another rain garden is a great option. You can also send it to another garden area or a storm drain. Above all, make sure water flows away from your foundation and your neighbors' property, unless you have decided to collaborate on a rain garden.

Under drains and dry wells

If you need overflow to go in a direction that gravity will not allow or your soils drain very slowly, choose an under drain or dry well. These options require more engineering, time, and materials to install, so we don't recommend using one unless it's absolutely necessary.

An under drain is composed of a drainpipe and gravel base beneath the rain garden. The drainpipe should extend upward to the desired pooling depth, but must be at least 2 inches (5 cm) lower than the inlet. The drainpipe leads down to a gravel base surrounding the connection to a buried pipe, which leads to another rain garden, vegetated area, or storm drain. Cover the top of the drainpipe with 0.25-inch (6.25-mm) hardware cloth or a prefabricated grate to keep debris out.

A dry well is a pit filled with rubble or cobble stone underneath the rain garden basin. The space between the rocks provides extra water-holding capacity. A dry well is built by digging a deep basin—from 24 to 36 inches (60 to 90 cm)—and filling it with cobble or rubble to 14

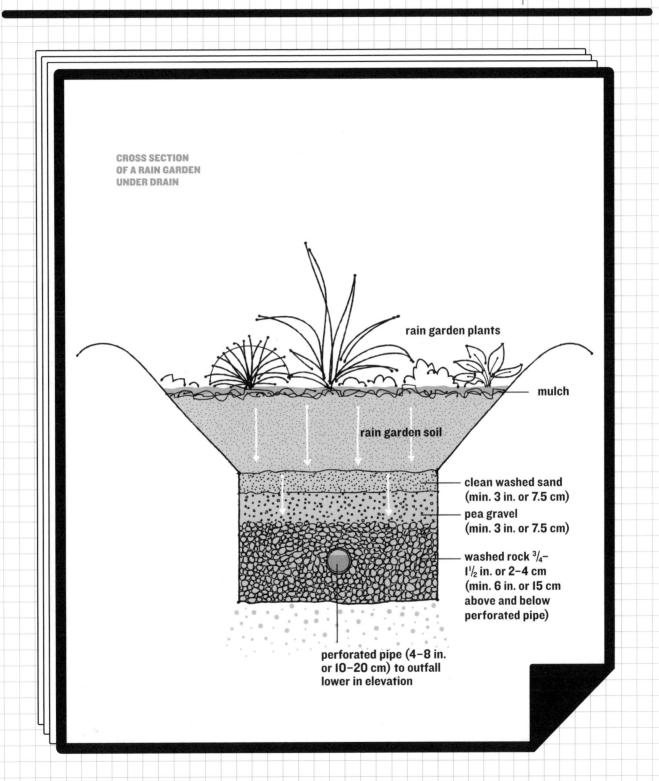

CROSS SECTION
OF A RAIN GARDEN
UNDER DRAIN

rain garden plants

mulch

rain garden soil

clean washed sand
(min. 3 in. or 7.5 cm)

pea gravel
(min. 3 in. or 7.5 cm)

washed rock $^3/_4$–
$1^1/_2$ in. or 2–4 cm
(min. 6 in. or 15 cm
above and below
perforated pipe)

perforated pipe (4–8 in.
or 10–20 cm) to outfall
lower in elevation

81

VARIOUS WAYS TO CAPTURE RAIN ON A STEEP SLOPE AND DETAIL OF A BOOMERANG BASIN

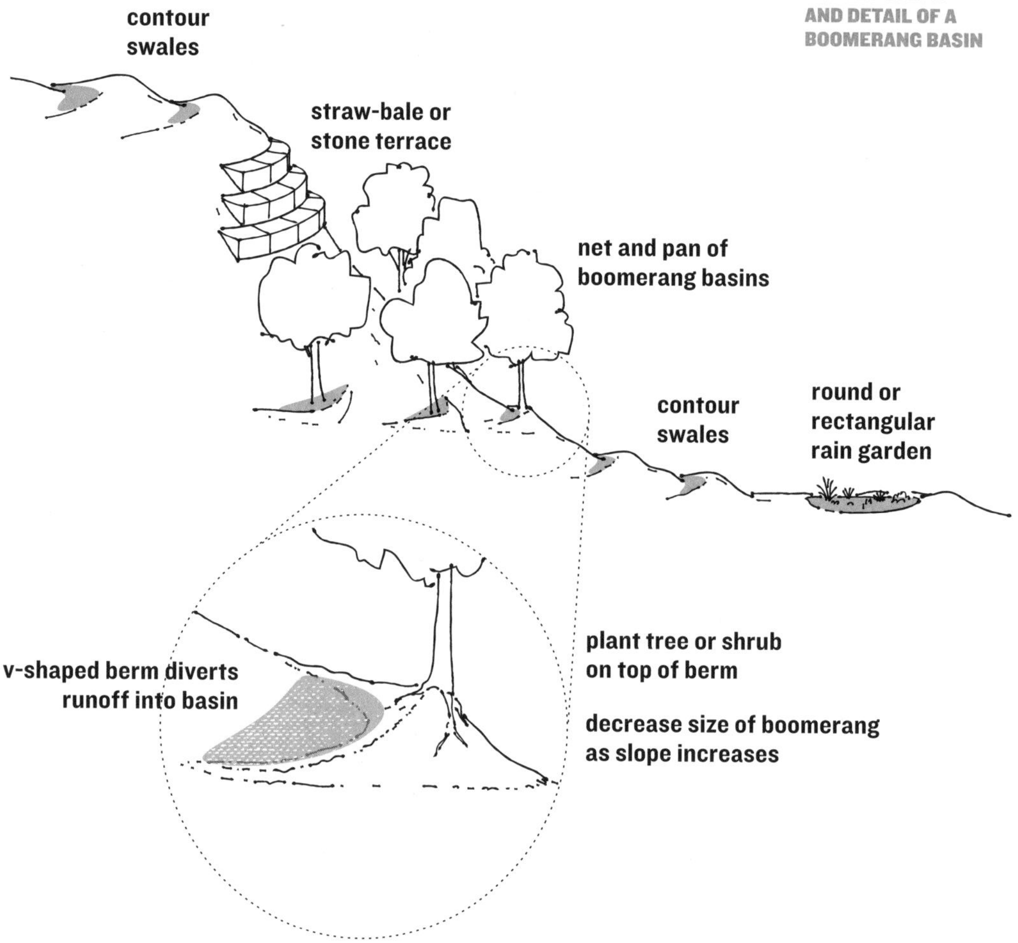

contour swales

straw-bale or stone terrace

net and pan of boomerang basins

contour swales

round or rectangular rain garden

v-shaped berm diverts runoff into basin

plant tree or shrub on top of berm

decrease size of boomerang as slope increases

boomerang basin detail
ground slopes between 18% and 30%

inches (35 cm) below grade. Next, 2 inches (5 cm) of gravel or river rock and then a 2-inch (5-cm) layer of pea gravel are spread on top. Finally, 10 inches (25 cm) of a well-draining soil (about 40 percent compost mixed with sand, pumice, or other locally available well-draining soil) is spread over the pea gravel. The basin is then topped with 3 inches (7.5 cm) of woodchips, bark mulch, or woody compost.

CATCHING RAIN ON STEEP SLOPES

If you live in a steep area prone to landslides, you're probably concerned about infiltrating more water into the ground. If your land is very unstable, you should seek out an engineer or soil scientist who's familiar with low-tech water harvesting and get his or her opinion. Even on the steepest sites, you can infiltrate some water—as long as you also plant vegetation that will transpire that water and stabilize the soil. Before we describe the techniques, here's a quick detour into the dynamics of soil on slopes.

The earth's surface is created by a constantly shifting dance between mountain building and erosion. As soon as plate tectonic forces thrust mountains upward, ice, wind, water, and microbial life begins to wear them away. Rock weathers to soil. Gravity pulls boulders, pebbles, and soil inexorably downhill. Soil slips from hilltops and accumulates on slopes, usually so slowly we perceive it only in the crooked line of an old fence or soil washed down a road cut.

The force of gravity usually keeps soil firmly attached to the rock beneath. In addition to gravity, friction between soil particles—known as the force of cohesion—keeps soil stuck to itself. Plant roots knit soil together very strongly, which is one reason that clearcut slopes often fail. In order for a land-

slide to occur, some force must push soil particles apart enough to overcome the forces of gravity and cohesion. This force is water pressure in the soil pores, which increases as soil becomes more saturated. At some point of saturation, water pressure along the plane where soil meets bedrock overcomes gravity and cohesion, and the soil slides down like a block down a lubricated slide. In general, the steeper the slope and the deeper the soil, the less saturated the soil must be before it slides.

When trying to control runoff on a slope, you want to infiltrate enough water to support plants with deep, fibrous roots—to help knit the soil together and make the soil less prone to slide—but you don't want to infiltrate so much water that the soil becomes saturated. The solution? Start very small. If water flows across the slope in a continuous sheet, dig hundred of pits that are just half a shovel deep, spread out across the landscape. These pits will trap organic matter and infiltrate some water without concentrating it where it could cause a slope failure. On more gradual slopes, dig small boomerang-shaped basins. Also, build long contour swales at the base and top of the hillside, where the ground slopes 15 percent or less.

If water is collecting into rills (small intermittent streams)—less than 1 foot (0.3 m) wide and 1 foot (0.3 m) deep—drive stout stakes along a contour line and weave branches between the stakes, making sure that the woven branches are close to the ground surface. Bundle small sticks and lay them on the ground uphill of the woven branches. This weir will slow flowing water, catch debris, and eventually cause the rill to fill with soil. Alternatively, dig level terraces, lay straw bales along them, and pin the bales into the hillside with wooden stakes. Make the weirs or bale terraces as long as you can and key them into an existing bush or clump of grass—otherwise water will cut around the end. These structures spread water across the slope and hold it in the brush or bales, creating nutrient-rich terraces where plants can establish. The plants then hold the soil and transpire water, thereby decreasing the risk of landslides. Watch the weirs during the rainy season, and repair them if they wash out.

PROJECT

HOW TO BUILD BOOMERANG BASINS ON SLOPES

Follow these directions to create a network of small, V-shaped basins along a hillside that slopes from 16 to 40 percent. The size of the basins depends on the slope. On 16 percent slopes, basins can be up to 6 feet (1.8 m) or more on a side and 18 inches (45 cm) deep. On 40 percent slopes, the basins should be no more than 2 feet (60 cm) on a side and 8 inches (20 cm) deep.

MATERIALS

PICKAXE   MATTOCK OR MCLEOD

1. Stand on the downhill side of a tree or shrub (or a place you will plant one). With a pickaxe or stick, draw a V uphill from the tree, so that the tree is in the point of the V.

2. Walk to the top of one of lines you just drew. Draw another V uphill from this point. Repeat for the other arm. The pattern of lines should start to look like a fishnet. Continue until the pattern covers the slope.

3. Dig basins in between each set of lines. Stand at the point of a V. With a mattock or McLeod (a firefighting tool), loosen soil and pull it toward the line. Make the berm largest near the point of the V, and taper it upslope. Tamp the berm well with the flat side of a large hoe or McLeod or by walking on it. If no tree or shrub is at that site, plant one on top of the berm.

LAST WORDS

CONGRATULATIONS, you've designed your rain garden. Now get ready for the real fun—getting your hands dirty and putting plants in the ground. In the next chapter, we explain how to trace swales and basins onto the ground, dig basins, lay pipe, and prepare the soil for plants by adding compost and mulch. As you move from design to construction, remember that rain gardens are forgiving. Don't worry if your basin ends up a little bit larger or smaller than you planned. A rain garden of any size will help your local creek, river, or bay and turn a corner of your garden into a buzzing, chirping, blossoming oasis.

WORSHEET

Rain Garden Information

RULE-OF-THUMB CALCULATIONS

TOTAL CATCHMENT AREA (from chapter 1):

....

EFFECTIVE RUNOFF AREA (optional):

....

| **CATCHMENT AREA DRAINING TO** | **BASIN 1:** | **BASIN 2:** | **BASIN 3:** |

....

| **TOTAL INFILTRATION AREA** (10% to 30% of catchment depending on soil type) **FOR** | **BASIN 1:** | **BASIN 2:** | **BASIN 3:** |

....

| **RAIN GARDEN DEPTH:** | **BASIN 1:** | **BASIN 2:** | **BASIN 3:** |

ENGINEER CALCULATIONS

DESIGN STORM RAINFALL INTENSITY: **TOTAL RUNOFF VOLUME:**

....

RAIN GARDEN DEPTH (based on soil type, from chapter 1):

....

| **AREA** | **BASIN 1:** | **BASIN 2:** | **BASIN 3:** |

....

PARTS CHECKLIST
(note all that apply to your design)

_____ Length of downspout or rain chain needed

_____ Length of runnel needed

_____ Length of 4-inch (10-cm) pipe needed (specify PVC or flexible drainpipe)

Pipe fittings needed (record number needed)

_____ couplings

_____ 90° elbow

_____ T fitting

_____ downspout adapter

Other materials needed

_____ tons river rock

_____cubic yards (cubic meters) amended soil

_____ cubic yards (cubic meters) mulch (specify kind)

_____ flagstone, brick, pavers, boulders etc. (specify quantity)

BUILDING YOUR RAIN GARDEN

IN THIS CHAPTER WE EXPLAIN

HOW TO TURN DRAWINGS AND FIGURES

into rain garden basins ready for planting. We discuss when to dig and plant, compare surveying and digging tools, and walk you through layout, excavation, and soil preparation for rain gardens of different sizes and shapes. We help you chart a path for rain to flow from your roof to the rain garden, whether your house has downspouts, rain chains, or sculptural channels. Being community-minded do-it-yourselfers, we encourage you to round up a group of friends and dig your rain garden by hand. However, we also explain how to dig a rain garden with an earth-moving machine.

WHEN TO DIG AND PLANT

PREVIOUS SPREAD: Large boulders accentuate this rain garden while acting as stepping stones through the garden.

OPPOSITE: This rain garden receives runoff from the patio and surrounding hardscapes.

Because building a rain garden involves both digging and planting, the best time of year to create a rain garden depends on your local climate. It's possible to dig any time of year, but it's much easier when the ground is not frozen, soggy, or baked into adobe. Most of North America and Europe receives winter precipitation, either as rain or snow, so a good time to work is usually in the spring after the soil has dried out somewhat or after the first couple of rains. If your soil is high in clay, digging when the soil is very wet will compact it and form slick layers, thereby decreasing the drainage rate of your rain garden. In wet climates, you may want to dig once the ground has dried somewhat, then plant during the next rainy season.

The best time to plant varies widely, especially if you want to minimize supplemental irrigation. Here are some things to keep in mind:

▶ Rain garden plants will require the least supplemental irrigation to become established if they're planted during a cool, rainy season. Depending on where you live, this may be winter, early spring, or late autumn.

▶ Newly transplanted plants lack developed root systems and are prone to heat and water stress, so avoid planting during very hot periods.

▶ Because most rain garden plants need to put down new roots before the ground freezes, you should avoid late autumn planting if winters are severe in your area. If in doubt, consult a local rain garden guide or your local agricultural extension service for recommended times to plant.

After digging the basin, increase infiltration by poking holes in the soil with a digging fork.

GETTING READY TO DIG

If you dig by hand, you may have most of the tools you need in your garden shed. At minimum, you'll need a pick, shovel or spade, digging fork, and a wheelbarrow to move excess dirt. A rock bar is useful if you have rocky soils. A rake or a McLeod (a firefighting tool) is handy for shaping and smoothing berms and basins. You will also need some way to check slopes and levels.

Drainpipes and diversion swales must slope at least 2 percent for water to flow down them, while the berms that surround rain garden basins must be level in order to contain pooled water. The rain garden inlet must be lower than the downspout, and the outlet must be lower than the inlet. Which tool you use to measure slopes and levels is a matter of personal prefer-

ence. If you have access to and know how to use a string or laser level, you can use these tools to level the top of a berm the same way you level a header or foundation form. If you don't own a laser level and want an inexpensive way to check levels quickly, you can make your own water level using clear vinyl tubing. (See the box, opposite)

Before you dig, get your local utility or a private company to mark buried gas and water lines. It usually takes 48 to 72 hours for locators to mark the lines. Utility companies will usually do this for free, but may only come up to the property line. If so, you will need to hire a private company to mark utilities on your property. Gas and water lines should be buried at least 18 inches (45 cm), but if the survey shows pipes crossing your rain garden site, strongly consider moving the rain garden. As you dig, watch out for irrigation lines, old clay sewer lines, and other miscellaneous pipes that the surveyors may have missed.

PROJECT

HOW TO BUILD A SIMPLE WATER LEVEL

Water seeks its own level. Ancient builders used this principle to level walls and foundations. You can make a simple water level by attaching a clear vinyl tube to two sticks, then filling it with water. You check levels with a ruler attached to each stick (or drawn with a permanent marker). The sticks don't need to be exactly the same length, but the numbers on the rulers should be the same height when both sticks are side by side on level ground.

MATERIALS

CLEAR VINYL TUBING, 0.5 INCH (1.3 CM) IN DIAMETER AND 25 FEET (7.5 M) LONG (OR LONGER IF YOU'RE DIGGING LONG CONTOUR SWALES)

TWO STRAIGHT STICKS OR POLES (USE SCRAP 1 × 2 INCH LUMBER OR BAMBOO), EACH APPROXIMATELY 5 FEET (1.5 M) LONG

EIGHT ZIP TIES OR 2 FEET (60 CM) OF WIRE

HAND SAW (TO CUT STICKS TO LENGTH)

PLIERS (IF USING WIRE)

TWO YARDSTICKS OR METER STICKS (OPTIONAL)

SHORT SCREWS AND SCREWDRIVER (OPTIONAL)

1. Using a zip tie or piece of wire, attach a yardstick (or meter stick) so it is flush with the top of each stick, or mark each stick in inches or centimeters using a permanent marker. Place the sticks side by side on level ground to make sure the numbers on the rulers line up when the bottoms of the sticks are lined up.

2. Attach one end of the vinyl tubing at the top of one stick with a zip tie or pieces of wire, and run the tubing along the entire length of the stick, attaching it at various points. Repeat with the other stick, so that the tubing forms a long U with the tops of the sticks attached to the open ends of the tube. The tubing should be at least 25 feet (7.5 m) long, but can be up to 100 feet (30 m) if your property is particularly large.

3. Fill the tubing with water. This is the trickiest part. Use a measuring cup, watering can, or funnel to pour water into one end of the tubing, or place one end in a bucket of water and suck hard on the other end, siphoning it.

4. Once the water reaches the ruler on the stick opposite the side being filled, get rid of any air bubbles by holding the sticks upright above your head. The air bubbles will slowly rise to the top.

USING THE WATER LEVEL

This is a two-person job, with each person holding one stick. One person (the anchor) stands at the end of the berm or future swale, marks the spot with a rock, stick, or surveyor's flag, and holds the stick on the marked spot as vertical as possible. The other person (the mover) walks several steps across the berm or slope. Each person looks at the height of water in the tube and calls out the value. The mover moves up or down the slope or berm until the water is at the same level on both rulers (for a level berm) or some specific distance higher on the mover's ruler (for a sloping trench or berm). For example, to lay out a diversion swale with a 2 percent slope, the mover would move 10 feet (3 m) away from the anchor, then move up and down slope until the water in the mover's tube is 2.5 inches (6 cm) higher than on the anchor's. Note that both people have to keep checking the height of the water because the water moves up and down in the tube as the mover moves up and down the slope. Once both sticks are at the correct level, the mover marks the spot at the bottom of the stick with a flag or stake and continues across the slope. Once the tubing is fully extended, the mover becomes the anchor, and the anchor leapfrogs along. At the end of this process, the flags will mark the path of the desired level or sloping berm.

You can use flour to mark the rain garden berm (outer line) and floor (inner line). A stick marks the inlet.

RAIN GARDEN LAYOUT

By now you're probably itching to move from layout to earthmoving. First, double check that the map you've made matches the site. Mark the place where you will dig the basin with a flag or potted plant. Then make sure that the rain garden area is lower than your collection area and that water can flow freely to the rain garden. Remember, the swale or pipe needs to slope at least 2 percent.

You can use marking paint, powdered chalk, stones, bamboo stakes, or survey flags to mark basin edges and pipe runs. Or you can use our favorite tool, a garden hose. A hose is easy to see and move around, and it makes smooth curves that are great for marking round or amoeba-shaped basins.

The instructions that follow will guide you through laying out basins and contour swales. As you lay out your rain garden, don't be afraid to tweak the design to make digging easier or to incorporate an existing plant or garden feature into the rain garden.

BASIN SHAPE CONSIDERATIONS

Rounded or rectangular basins are appropriate for sites that slope up to 15 percent. The classic rain garden is wide and roughly round, often with amoeba-like extensions. There's nothing wrong with straight lines in a rain garden, but rectangles have much less edge than shapes with indentations or extensions. If your rain garden is in a rectangular courtyard or along the curb, consider a more complex shape for the basin to create more visual interest. Complex shapes increase the length of the edge relative to the area of the basin, and edges and varied slopes create more microhabitats for plants and wildlife.

On hillsides that slope between 5 and 20 percent, you can build contour swales, long skinny basins that spread water across the landscape. On gentle slopes, dig a contour swale if you want to plant a row of trees along the berm or for aesthetic purposes. On slopes between 15 and 20 percent, use contour swales to slow erosion and spread water horizontally along a slope, because concentrating it in one place could trigger a landslide. It's a good idea to dig two or three shallow contour swales that progressively overflow into one another down slope. This strategy increases the amount of water that can be stored in the soil. As a rule of thumb, space swales 20 times the basin depth: for example, 20 feet (6 m) apart for a 1-foot-deep (0.3-m) swale.

HOW TO LAY OUT A CLASSIC RAIN GARDEN

On flat sites, simply lay out an interesting shape with a garden hose or powdered chalk. On sloping sites you can either use fill from the basin to form the berm or dig farther back into the hillside and use native soil for the berm. Either way, make sure the berm will be level, and then mark the outer edge with a hose or powdered chalk.

Once you've outlined each basin, step back and try to picture a rain garden there. Does the basin seem too large or too small for the space? How does the shape interact with the plantings and structures around it? Again, play around with the shape and size until you're satisfied, and mark the basin outline.

2–8% slope

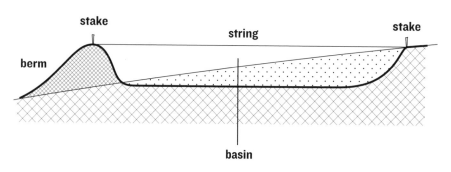

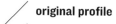

9–15% slope

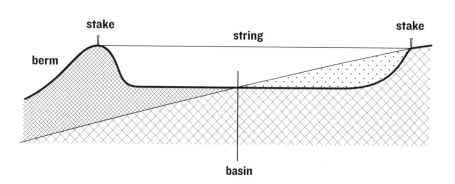

CLASSIC RAIN GARDEN basins are laid out differently on gentle and moderate slopes. Place stakes at the downhill edge of the rain garden and run level strings to the top edge. These strings mark the level of the berm. On gentle slopes (up to 8 percent), the berm sits on native soil. On moderate slopes (9 to 15 percent) use fill excavated from the uphill side of the basin to form the downhill side of the basin and the berm.

The basin sides should have a 30° (approximately 60 percent) slope, otherwise the sides may cave in or wash down into the basin. To calculate the necessary basin inset, multiply the basin depth by two. For example, a basin that is 9 inches (22.5 cm) deep would have an 18-inch (45-cm) inset, so you would mark a second line 18 inches (45 cm) in from the basin outline. This inner line marks the bottom of the basin.

Mark the place where water will flow into the rain garden with a flag or rock. Decide where you want the water to overflow if the rain garden fills up, and mark that too. Make sure that the overflow will not flood your basement (or a neighbor's property or basement).

HOW TO LAY OUT CONTOUR SWALES

When building a contour swale, use a level of your choice to survey contour lines, which are similar to the lines on a topographic map. A contour line is level, that is, it is the same elevation everywhere. You'll need enough surveyor's flags or 18-inch (45-cm) sticks to mark off the entire swale length at 6-foot (1.8 m) intervals.

Stand where the pipe or diversion swale will enter the contour swale. Place a surveyor's flag or 18-inch (45-cm) stick in the ground at this point. Use a water level or laser level to survey a contour line across the slope. Place a flag or stick into the ground every 6 feet (1.8 m) or so along the future swale. When you dig the

THE SIZE OF THE CONTOUR swale basin and berm depend on slope. The steeper the slope, the longer and narrower the swale should be. The basin of the swale should be at least 18 inches (45 cm) wide along contour and 12 inches (30 cm) deep. The berm is a mirror of the basin and is composed of soil dug from the basin and piled along a contour down slope. The berm should be the same width as the basin and as tall as the basin is deep.

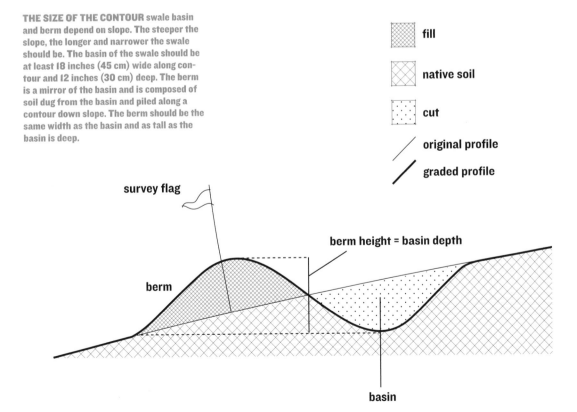

fill

native soil

cut

original profile

graded profile

survey flag

berm height = basin depth

berm

basin

swale, the flags will stick up through the top of the berm, so make sure they're at least 12 inches (30 cm) above ground. If possible, end your line at a tree, bush, or large rock, so that water flowing down the slope past the end of the berm won't wash it away.

Now stand back and imagine that the line of flags is the berm of the swale and the basin is just uphill from the berm. The basin will be at least 18 inches (45 cm) wide and 12 inches (30 cm) deep. The berm is a mirror of the basin, because it's made from the dirt dug out of the swale. If the contour swale seems like it's in the wrong place, move it uphill or downhill. Then decide where you want the water to overflow if the swale fills up, and mark that as well.

CHECK THE BASIN SURFACE AREA AGAINST THE DESIGN

The last step before digging is to make sure the basin is the right size. Measure the length and width of the bottom of your basin (that is, the shape of the inset), and multiply these values to calculate the area. (Even if your basin is very irregular, you can assume it is roughly rectangular or circular.) If the basin is much larger or smaller than the area you calculated in the design, adjust the layout accordingly by moving the hose or chalk mark.

LAYING OUT THE CONVEYANCE SYSTEM

Now that you've laid out your rain garden, look across the landscape to the roof or driveway that is your catchment surface. Refer to the flow map you made during the site assessment to help you visualize how water will flow there. What is the easiest way to move the water to the rain garden site? You can use diversion swales, buried pipe, channels, runnels, or a combination of strategies, or simply allow water to flow across the surface in gently sloping, vegetated areas. All parts of the conveyance system need to slope gently downhill.

DIVERSION SWALE LAYOUT

Diversion swales can be straight or curve in gentle arcs, and they are easy to shape as you dig. They must slope gently (between 2 and 8 percent). If the slope is less than 2 percent, water will not flow, and if greater than 8 percent, flowing water will erode soil that will eventually fill the rain garden basin. Don't worry about the exact slope at this stage, but do choose a path that slopes downward as gradually as possible. The closer the path is to the final swale slope, the less you'll have to dig later.

Play around with swale layout until you've created a dynamic and functional waterway from downspout to basin. If your plan calls for a series of basins, stagger the inflows and overflows so that water moves in a sinuous zigzag across the landscape.

BURIED PIPE LAYOUT

If you plan to move water in a buried pipe, choose the shortest path between the downspout and rain garden. Buried pipes can slope from 2 percent to vertical. Mark a straight line from the downspout to the basin with a hose or powdered chalk. If the line crosses a mound, rise, or obstacle like a tree root zone, move the line—avoiding such obstacles could save you hours of digging. Make sure that you can dig a trench along the line that slopes downhill.

If you do need to divert from a straight line, try to make all bends less than 90°. Tighter turns tend to clog and decrease flow through the pipe. For corrugated pipe, lay out curving trenches. For rigid pipe, use 45° and 22° elbows if possible.

DIRECTING WATER FLOW ACROSS HARDSCAPE

If a pipe or diversion swale will cross a cement walkway or patio, you will need to cut through

LAYING OUT
AND DIGGING A
CONTOUR SWALE

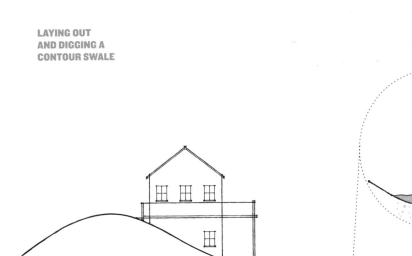

finished swale
two years later

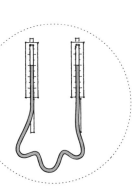

water at the same level
on both measuring
sticks means the
contour is level

finding contour lines
with a water level

marking contour
lines with flags

digging the swale

RAIN GARDEN CONSTRUCTION CHECKLIST

Before you dig, check to make sure that:

- [ ] Rain will flow toward your rain garden.
- [ ] Basin edges and floors are marked, and the bottom of the basin is inset twice the depth on all sides.
- [ ] The marked basin is as large as the design specifies.
- [ ] Each basin has an overflow drain marked that leads to another basin, a well-drained area, or a storm drain.
- [ ] Buried pipes and diversion swales are marked, and obstacles like roots and mounds have been avoided.
- [ ] Diversion swales slope between 2 and 8 percent. Steeper swales will need to include check dams to control erosion.
- [ ] The location of shallow utilities is marked.
- [ ] Compost, mulch, river rock, and tools are on hand.
- [ ] You know where you want to put the excavated soil.

OPPOSITE:
A stacked-stone sitting wall with vegetation filling the interstices, an example of a contour terrace.

that hardscape with a gas-powered demolition saw. Make two lines with grease pencil or chalk, approximately 2 inches (5 cm) wider than the drainpipe or runnel, across the hardscape. A cut through a walkway can be filled back in with smooth river rock and a stepping stone or covered with a metal grate.

DIGGING RAIN GARDEN BASINS, SWALES, AND TRENCHES

Once you have your basin and conveyance laid out, you're almost ready to dig. The last point to consider is where you will put the soil that you excavate. If your rain garden will hold 500 gallons (2 cubic meters) of water, you'll need to find a place to put that much of dirt. Will you fill a low spot, build a large berm or mound, use the soil to make adobe bricks or a cob oven, or put it on a tarp with a Free Dirt sign? Figure this out before you start digging.

Most landscapers we interviewed for this book dig rain gardens using machines. If your soil is very hard or rocky, you may decide to go this route as well. But digging your basin by hand saves money, reduces greenhouse gas emissions, and has several other advantages. Digging is good exercise and it gives you an up-close understanding of your soil. Digging can be a lot of fun for a group of friends if you pitch it as a skillshare. Kids are usually avid and persistent diggers—even small ones are great at stomping down berms, raking, or picking out stones. And hand digging makes it easier to till as you go, increasing the soil's ability to infiltrate water. If digging a giant basin by hand seems daunting, consider digging several small ones, one at a time.

Digging a trench and checking that the pipe slopes downhill.

TIPS FOR DIGGING WITH HEAVY EQUIPMENT

Because heavy equipment compacts soil and reduces its ability to infiltrate water, never drive a backhoe or excavator over or into the rain garden. Instead, park it outside the line marking the edge of the basin.

Unless you already own a backhoe, rent the smallest one that will do the job. A small excavator will work, unless the soil is very rocky or your rain garden is very large. Before you call the rental yard, measure the longest distance from the edge of the rain garden to the center. Then make sure the equipment you rent has a digging arm that is long enough to reach.

Once you have finished digging the basin to the right depth, scratch the bottom and sides with the backhoe teeth to loosen the soil. Then finish the basin by hand with a shovel, fork, and rake.

TIPS FOR DIGGING BY HAND

For a certain type of gardener, nothing is as satisfying as a hard, sweaty digging job. The task is clear, the work straightforward, and the cold drink at the end a fitting reward. If you're this type of gardener and have evolved techniques that suit your soil and temperament, feel free to skip our digging suggestions. Otherwise, read on.

Digging should be athletic but not exhausting—if the task becomes tiring or tedious, rent a rototiller, call a friend for help, or hire a few laborers to finish the job. Don't rush the job, dig through a heat wave, or ignore aches and pains.

As you dig, loosen the soil with a digging fork, then scoop it out with a shovel. Use a pickaxe or mattock to loosen hard clay and to pry out rocks. If you're digging in clay or loam, shoveling and foot traffic will compact the soil, so when the basin is as deep as you want it, loosen the soil at the bottom with a digging fork. The more holes you poke in the bottom of the basin with a digging fork, the better the water will infiltrate.

THROW A PARTY TO LIGHTEN THE WORK

TELL ALL YOUR FRIENDS you are making a rain garden. As you're doing the design and research, you can get friends involved by asking them along on hikes to local natural areas to check out water-loving plants, going to native plant sales together, and asking for cuttings or divisions of rain garden plants they might have growing in their gardens.

Invite a lot of people to your digging party. You'll want at least six, and if fifteen come the work will only be easier. Bill the day as a skillshare—you will share what you have learned about rain gardens and they will learn how to dig one. Tell them you'll provide lunch and beverages. Mentioning home-canned jam, fresh garden produce, or homemade beer or ice cream always helps draw a crowd. You can also hold speed or endurance competitions for teams of diggers, with a prize for the winning team.

Spend an afternoon before the workday preparing the site by marking where swales, buried pipes, and basins will go. Read through this chapter carefully, and think about how you will explain the digging process to your helpers. Make a list of tasks on a large piece of paper and post it near the work site. Transplant any plants to be removed from the rain garden area. Gather all the necessary materials, the pipe, fittings, compost, mulch, rock, and plants. Borrow shovels, pickaxes, rakes, and wheelbarrows from friends or your local tool lending library.

On the day of the work party, have work gloves, tools, water, shade, and rain garden resources available. Explain the day's work, and then divide people into basin digging, downspout plumbing, and planting crews. If people show up late, ask these crews to fill them in—people learn best by teaching others.

Take a long break for lunch, and stick to your ending time. Plan for 5 hours, and aim to finish early. Above all, make sure the work is fun and that people learn how and why to make a rain garden. Let everyone know that when they're ready to dig a rain garden, you'll come over to help.

STEP-BY-STEP DIGGING INSTRUCTIONS

The instructions that follow will guide you through digging round or rectangular basins and long, skinny contour swales. After the basic instructions, we discuss modifications for different drainage conditions. The following considerations apply to all basins:

▶ When the basin is the size and shape that you want, step back and look it over. Does it seem too small or too large for the space? Is the shape interesting? Trust your intuition, and don't be chained to your site plan. Play around with the shape of the basin until it looks good with the rest of your landscape.

▶ Once you've dug the rain garden basin, dig the swale or trench for buried pipe from the downspout to the rain garden inlet. Check that the trench slopes down from the downspout to the rain garden, then bury drainpipe if you're using it.

▶ Dig the basin outlet last. Remember that the outlet must be lower than the inlet. How much lower? Opinions vary, but 2 to 6 inches (5 to 15 cm) is a good rule of thumb.

▶ If you're using an overflow pipe, the bottom of the pipe should be about 3 inches (7.5 cm) lower than the bottom of the inlet. The top of the pipe can be level with the berm if you like.

▶ If the outlet leads to a diversion swale or second basin, make a depression for the outlet that is 4 inches (10 cm) lower than the berm, then line the outlet with stone or brick.

▶ If you're diverting your roof runoff to a series of basins, dig the highest basin first and then dig the basin below. Connect the two basins with a swale or buried pipe the same way you connected the downspout to the basin.

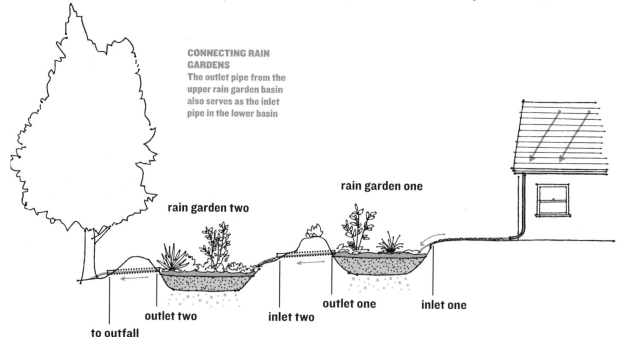

CONNECTING RAIN GARDENS
The outlet pipe from the upper rain garden basin also serves as the inlet pipe in the lower basin

rain garden two

rain garden one

outlet two

to outfall of lower elevation

inlet two

outlet one

inlet one

DIGGING A CLASSIC RAIN GARDEN BASIN

Don't build a classic basin on slopes steeper than 15 percent, which could cause a landslide. In these cases, dig a contour swale instead.

1. Dig down to the designed depth starting at the inner line you marked, and excavate the center of the basin to this depth.

2. Cut the edges of the basin back to the outer line you marked. The sides should slope at about 60 percent (30°)—any steeper and the sides will slump into the basin.

3. On flat sites, use the excavated soil to make a berm all the way around the basin. If you're digging on a gentle slope, place the soil on the downhill side of the basin to make a berm. Compact the berm by stomping on it so that water rushing into the rain garden won't erode it.

4. Level the bottom of the basin. This is a critical stage in the construction of the rain garden. The basin area was designed to drain a certain amount of water per day. If the bottom is not level, water will not drain as effectively.

5. In the place where you want the basin to overflow, make the berm 4 inches (10 cm) lower to create the outlet. Build a spillway by laying cobble, brick, or rubble around the overflow point. Alternatively, you can put a 4-inch (10-cm) drainpipe in the spillway, cover it with dirt, and set cobble or river rock around the pipe to conceal it and protect the surrounding soil.

Soil amendment for well-draining soil

If your soil is loose and sandy or loamy and infiltrates 1 inch (2.5 cm) or more per hour, spread 3 inches (7.5 cm) of woody compost on the bottom and sides of the basin. Using a digging fork, loosen the soil to a depth of 6 inches (15 cm) and mix the compost into the soil. Don't worry about mixing it in completely.

Soil amendment for poorly draining soil

One option for dealing with poorly draining soil is to fill the basin with looser soil. Make well-draining soil by mixing 40 percent com-

OPPOSITE:
A variety of depths and shapes create microhabitats for an array of plants and animals.

post with 60 percent sand, pumice, or other fast-draining material. Till this mixture into a minimum of 3 inches (7.5 cm) of native soil at the bottom of the basin, and then add the soil mixture 6 inches (15 cm) at a time, compacting each layer by walking on it. Don't fill the basin completely—leave the space for pooling that you calculated in the design chapter. Top with 3 inches (7.5 cm) of woodchips, bark mulch, or woody compost.

If your soil drains less than 0.1 inch (0.25 cm) per hour, install a dry well under the rain garden basin. Dig a deep pit in the bottom of the basin (24 to 36 inches; 60 to 90 cm), and then fill the basin half way with cobble or rubble. Fill the remaining depth with a thin layer of pea gravel, finished by a thick layer of amended soil (40 percent compost, 60 percent sand, pumice, or other locally available drainage material) until almost flush with the basin floor. Cover the top of the dry well pit and basin bottom with at least 3 inches (7.5 cm) of woodchips, bark mulch, or woody compost.

DIGGING A CONTOUR SWALE

Contour swales are designed to infiltrate water and are level, whereas diversion swales are designed to move water across the landscape and slope gradually, generally around 2 percent.

1. Using a laser level or water level, mark a contour line with flags. Transplant any plants growing along the line that you want to save.

2. Choose your tool. A mattock (the wide end of some pickaxes) or a McLeod (a firefighting tool) will do the job. A rake or wide-bladed hoe is useful for smoothing the finished surface.

3. Stand just downhill from the line of flags. While facing uphill, aim the bladed end of your tool 18 inches (45 cm) uphill from the line of flags. Swing it down hard to loosen the dirt. Then scrape the loose dirt back toward you and pile it along the contour line. (If you're using a mattock, turn the blade sideways and scrape up dirt with the side of the blade.) Repeat this motion until you have a wide,

shallow depression uphill of a wide, low berm.

4. Continue along the slope until the berm and basin extend the length of the surveyed contour line. You should have a line of flags (or sticks) sticking out the top of the berm. Make sure the berm is centered along the line of flags. Using your feet, tamp the berm firmly. Then check the level of the top of the berm with a water level to make sure it is at the same elevation along its entire length. Add dirt to low spots and remove it from bumps. Note that the berm of the swale must be on contour, but the basin elevation can vary with plants' water needs. For example, you can dig deeper pits in the swale near water-loving trees to give them extra water.

5. Walk to one end of the swale. Do you see a tree, bush, or rock nearby and slightly uphill? Put a flag or two in between the end of the berm and this tree or rock. (If there is no tree or rock, you can plant a bush or tree later.) Now continue digging the swale along this line (it should curve slightly uphill). You want the top of the berm to remain level and to key into a tree or rock. The basin will slope downhill. Keying the swale in this way keeps water from cutting around the end and eroding the berm. Repeat at the other end of the swale.

6. Decide where you want the swale to overflow, and make the berm 4 inches (10 cm) lower in this spot. Build a spillway by laying cobble, brick, or rubble around the dip.

7. Mulch the berm and basin with a layer of straw, pecan hulls, bark, or woodchips.

HOW TO DIG DIVERSION SWALES AND BURY PIPE

Once you've dug the rain garden basins, the next step is to build diversion swales or buried pipes to carry runoff there. We present separate instructions for diversion swales and buried pipes, but you can also combine the

PROJECT

HOW TO BUILD A CHECK DAM

A check dam slows the flow of water running down a diversion swale. Check dams are usually made with rock, but can also be made by weaving sticks around wooden stakes. A rock check dam looks like a stone step above a flat stone landing that acts as a splashguard. Water can flow through the gaps between the rocks and sticks, but soil gets trapped behind the check dam. The splashguard breaks the force of water that flows over the check dam during heavy rains. Build the check dam perpendicular to the diversion swale, in a straight section of the swale.

For a 3-foot (0.9-m) wide swale, you'll need five to seven blocky rocks for the check dam and between five and fifteen flatter rocks for the splashguard (depending on the size of the rocks). Choose rocks that are about 3 inches (7.5 cm) taller than the depth of the diversion swale. For the splashguard, use thinner, flatter rocks.

MATERIALS

ROCKS (ROUGHLY HEAD SIZED)
PICKAXE OR MATTOCK
SHOVEL
TROWEL
RAKE

1. Mark a straight line across the diversion swale that extends 1 foot (0.3 m) into the bank on each side.

2. Dig a trench 6 to 12 inches (15 to 30 cm) wide along this line. The bottom of the trench should be level and 2 to 4 inches (5 to 10 cm) below the bottom of the diversion swale. It is important that the trench extends into the bank of the diversion swale—otherwise, water will cut around the end of the check dam.

3. Lay one course of stones in the trench. The gaps between stones should be less than 2 inches (5 cm). Make the top of the check dam roughly level, but 1 to 2 inches (2.5 to 5 cm) lower in the center. Use your trowel to dig out under taller rocks and to shove dirt under loose rocks to set them.

5. Lay a splashguard of flat stones or river rock below the check dam.

6. Fill the trench with soil to the edges of the rocks, and smooth out the grades with a rake.

7. Pile small sticks or brush behind the check dam to catch debris.

two. For example, you can transition from swale to buried pipe to divert water under a path. Make sure to protect the soil around the pipe inlet and outlet with cobble stone or gravel.

DIGGING A DIVERSION SWALE

Diversion swales move water across the surface of the landscape, rather than hiding it beneath. They offer several advantages that pipes do not. Vegetated swales can provide habitat and food source for wildlife. They increase the amount of water you can infiltrate into your garden, because some water seeps into the ground en route to the rain garden. They are easy to maintain, because any debris that accumulates is easy to spot and remove. And the water draws the eye and soothes the mind as it splashes down the swale.

1. Dig a shallow, gently sloping depression from the rain chain or downspout to the rain garden basin.

2. Check that the swale slopes at least 2 percent and no more than 8 percent. Use check dams (low rock walls) to move water across steeper drops.

3. Where the swale enters the rain garden, cover the soil with river rock, cobble, or brick or concrete rubble.

4. Plant the swale with grass, wildflowers, or drought-tolerant emergents like sedges or rushes, or line it with river rock.

PROJECT

CONNECTING A DOWNSPOUT TO A DIVERSION SWALE

MATERIALS

DOWNSPOUT
(PVC OR METAL)

PVC SAW
(FOR CUTTING
PVC DOWNSPOUT)

HACKSAW (FOR CUTTING
METAL DOWNSPOUTS)

MEASURING TAPE

SHOVEL OR TROWEL

COBBLE STONE OR GRAVEL
(APPROXIMATELY 5 GALLONS,
0.02 CUBIC METERS)

PAVERS OR LARGE, FLAT
STONE

1. Install a downspout if none is present or cut the existing downspout to about 8 inches (20 cm) above the ground level. Attach a curved piece of downspout to direct water away from the foundation with the appropriate fitting. The downspout should extend 2 feet (0.6 m) away from the foundation.

2. Make sure that the ground slopes away from the foundation and into the swale—you don't want water pooling against the foundation. With the shovel or trowel, excavate a shallow hole under the end of the downspout piece to hold the flagstone, cobble, or pavers. (You can use several pavers, a circular pad of cobbles, or a flagstone surrounded by cobbles—just make sure that water will splash onto rock and not onto soil.) The hole should be round or square, 2 feet (60 cm) wide, and about 3 inches (7.5 cm) deep.

3. Lay the flagstone, pavers, or cobbles under the downspout, and backfill with dirt if necessary.

PLACING A BURIED PIPE

You can use either PVC or corrugated pipe to bring water to your rain garden. Use PVC on shallow slopes (2 to 4 percent), because the ridges in corrugated pipe reduce water flow.

1. Dig a continuously downward sloping trench from the bottom of the downspout to the rain garden. The trench should be 8 inches (20 cm) deep and at least 6 inches (15 cm) wide.

2. Lay 4-inch (10-cm) drainpipe in the trench and check that the pipe slopes at least 2 percent. Corrugated drainpipe can be bent to make turns, and these curves are less likely to clog than 90° elbows. Cut corrugated pipe with a utility knife. Use 90° elbows where tight turns are unavoidable. Rigid drainpipe can be cut with a hacksaw and connected with fittings, which come in 90°, 45°, and 22° angles. Gluing drainpipe is not necessary, although some plumbers prefer to use glue, especially in frost-prone areas where repeated freezing and thawing can pry pipes from their fittings.

3. Backfill the trench near the rain garden and tamp the dirt down by walking on it. (You will backfill the rest of the trench after you hook up the pipe to the downspout and make sure it slopes toward the rain garden.)

4. Place cobble stones, bricks, or rubble around, but not blocking, the end of the pipe where it flows into the rain garden. Use river rock, brick, decorative pavers, or flagstone to make a splashguard below the pipe to break the force of water flowing into the basin.

HOOKING UP TO YOUR RAIN GARDEN

When you designed the conveyance system, you chose to use downspouts or rain chains to move water from your roof to the ground and diversion swales or buried pipes to move water

Rain cups
connecting to
a buried pipe

Buried pipe connects the downspout in the background to this finished rain garden, it's entrance covered by river rock on the far right to prevent erosion.

into the rain garden. Here we describe how to hook up the conveyance system that will connect your roof to the rain garden. We assume you have already installed a diversion swale or buried pipe from the edge of the building to the rain garden basin.

CONNECTING DOWNSPOUTS

Downspouts can be made of galvanized steel, PVC plastic, or copper and are either rectangular or round in cross section. Four-inch (10-cm) drainpipe is round and made from corrugated black plastic or rigid white PVC. You can find fittings to connect either square or round downspouts to either corrugated or rigid pipe at a building supply store.

CONNECTING RAIN CHAINS

A rain chain can connect to a gutter or a canale and can be made of a variety of materials. Steel chain, decorative copper rain chain, or linked copper cups will all guide water from the roof to the ground. Buy chain 18 inches (45 cm) longer than you need. Copper rain chain and cups come with an expandable hanger that attaches to the gutter. If you're using steel chain, you can buy the expandable hanger at a garden store or fabricate a hanger from hooks or other hardware. You'll need to drill a hole in the gutter to admit the rain chain. Make sure the gutter is attached firmly and is sturdy enough to support the weight of the chain and water, then hang the chain and stake the bottom end to the ground with a landscape staple.

A splashguard of flagstone, pavers, or cobbles breaks the force of water running down the rain chain; angle the splashguard slightly away from the foundation. If the rain chain connects to a swale, the water simply runs across the splashguard into the swale. If the

PROJECT

CONNECTING A DOWNSPOUT TO A BURIED PIPE

MATERIALS

PVC SAW (FOR CUTTING PVC DOWNSPOUT)

HACKSAW (FOR CUTTING METAL DOWNSPOUTS)

MEASURING TAPE

SHOVEL OR TROWEL

DOWNSPOUT

4-INCH (10-CM) DRAINPIPE

90° ELBOW FITTING (SAME MATERIAL AS BURIED PIPE)

SHORT PIECE OF DRAINPIPE (SAME MATERIAL AS BURIED PIPE)

DOWNSPOUT-TO-DRAINPIPE ADAPTER FITTING (SAME MATERIAL AS BURIED PIPE)

1. Install a downspout if none is present, or cut the existing downspout to about 8 inches (20 cm) above the exposed drainpipe, which should be lying in the bottom of the trench.

2. Push the downspout adapter onto the bottom of the downspout. Then push the 90° elbow fitting onto the end of the drainpipe that runs to the rain garden. The downspout adapter should be 4 to 6 inches (10 to 15 cm) above the 90° elbow resting on the bottom of the trench.

3. Measure from the inside of the downspout adapter to the inside of the top of the 90° elbow fitting. Cut a piece of drainpipe to this length and fit it into the elbow, then attach it to the downspout adapter. You don't need glue because gravity will hold the drainpipe in place. (If this connection does start to leak during a rainstorm, you probably have a clog downstream in your system.)

4. Backfill the trench with dirt to cover the pipe, and tamp the dirt down by walking on it.

III

rain chain connects to a buried pipe, the splashguard conceals a buried cobble-filled 5-gallon (20-L) bucket that collects the falling water and connects to the buried pipe. Cut a hole in the side of the bucket near the bottom that is just barely larger than the outside diameter of the pipe, then slide the pipe into the bucket. Don't worry about caulking the connection.

USING SWALES BELOW GUTTERLESS EAVES AND DRIVEWAYS

OPPOSITE:
A cut through a walkway to allow water to flow to a rain garden can be filled back in with smooth river rock and a stepping stone.

In general, you want gutters on your roof to protect the building from falling rain. Older buildings often settle into the ground, which can cause pooling at the foundation, leading to flooding or mold problems. If your roof doesn't have gutters, it is crucial to grade the dirt around the building so that the ground slopes away from the foundation for at least 2 feet (60 cm). Then lay out and dig a diversion swale along the length of the building.

1. Dig a diversion swale that is 18 to 24 inches (45 to 60 cm) wide and about 12 inches (30 cm) deep. The inner edge of the diversion swale should lie 6 inches (15 cm) inside the drip line of the eaves or at the edge of the driveway or patio.

2. Place a layer of river rock or flagstone directly under this drip line to keep soil from washing away. Plant groundcover or a grass and wildflower mix in the swale, and mulch with 2 inches (5 cm) of woodchips or coarse compost.

3. The next time it rains, go outside and make sure the rain drips from the roof or flows from the patio or driveway and lands on the layer of rock to ensure the soil does not erode and slip away.

TESTING YOUR RAIN GARDEN

The last step before amending the soil is to make sure the rain garden will fill up, overflow, and drain properly.

☐ After adding compost or other amendments, will there be room for the amount of pooling you figured in the rain garden design? Refer back to your design notes for the depth of amended soils (if any), and plan on 3 inches (7.5 cm) of mulch.

☐ Can water flow freely from the inlet pipe to the outlet point?

☐ Is the berm level, except where you want the water to overflow? The outlet should be at least 2 inches (5 cm) lower than the inlet pipe or swale. Check the elevations with a string level, laser level, or water level, and tweak the height of the outlet or berm if you need to.

☐ Is the inlet and outlet protected from erosion by surrounding it with river rock, cobble stones, or broken brick or concrete? Throughout the system, add cobble or flagstone anywhere that water falls or flows quickly—under rain chains, below inlets, or on steeper slopes.

This is a good time to do a flow test. Unless it's already raining, put a hose at the downspout, turn it on, and let water flow down the pipe or channel into the rain garden. Vary the hose pressure to simulate a drizzle and a downpour. Then watch how the water flows. Is there anywhere that it washes soil away? Does it spread out evenly over the bottom of the basin? Once the basin is full, does it spill over at the overflow point without backing up into the inflow pipe or swale? If necessary, adjust the height of the outlet or add more stone.

NOTES ON AMENDING SOIL

If your soil drains well, all you need to do is add 3 inches (7.5 cm) of compost and mix it into the top 6 inches (15 cm) of soil in the basin. If your soil drains slowly, say less than 0.5 inch (1.3 cm) per hour, you'll need to take further steps if you want to infiltrate all the rain that falls on your roof.

Rain gardens are fairly cutting edge, which means that there's no clear consensus on the best rain garden strategies for difficult situations. Like edible and ornamental gardens, what makes the best rain garden is site specific and subjective. For instance, consider infiltration. According to the Oregon rain garden guide, where soils drain less than 0.5 inch (1.3 cm) per hour, it is recommended to dig extra-large basins and fill them with sandy soil. In the age of malarial sewage treatment plants and West Nile outbreaks, municipalities worry that slow-draining rain gardens will create mosquito problems, and so they err on the side of caution when it comes to drainage. But filling a basin with sandy soil isn't the only way to keep your rain garden from becoming a malarial swamp.

When Brad Lancaster removed the driveway of his Tucson, Arizona, home and replaced it with a basin to harvest runoff, the alkaline desert soil barely drained. After the first monsoon rain, the basin took 12 hours to drain. But when he planted saltbush (a Sonoran Desert shrub) in the basin, drainage improved. When the monsoon returned the next summer, the basin filled just as quickly but drained in less than 2 hours. The lesson? Rain gardens are living structures. Once plant roots punch thousands of tunnels into the soil, even the most compacted clay or caliche soils can drain well enough to avoid mosquito problems and to catch a lot more rain during a several-day storm.

What to do if your soil doesn't drain for weeks? First, make sure to build your rain garden at least 10 feet (3 m) from any foundation. Your property likely already has moisture problems from a high water table, and you don't want to make them worse. Make sure your rain garden has a good overflow, leading to the street or sewer if necessary.

In this case you could think of your rain garden as an ephemeral pond or vernal pool. Once common in seasonally wet grasslands, vernal pools fill up during the rainy season and then slowly dry out in the summer heat. Vernal pools are under threat from development and agriculture, so if you live in a place where they once existed, consider incorporating one into your landscape. Dig an extra-large basin, mix in 3 inches (7.5 cm) of compost, and add plants that can handle being underwater for a long time, such as sedge (*Carex*) and cranberry (*Vaccinium macrocarpon*). If your pond is wet for several weeks at a time, dragonflies will find it, and their larvae are voracious predators of mosquito

larvae. You could also build a bat or swallow box to encourage these mosquito-eating machines to take up residence.

Or you might follow the Puget Sound and Oregon rain guides' advice and think of your rain garden as a bog, a moist area with occasional standing water. Dig the basin 18 to 24 inches (45 to 60 cm) deep, and fill it partway with a quick-draining mix of compost and sandy soil (or pumice, sand, or other locally available drainage material). Rain will fill the basin but remain under the surface of the sandy layer except just after a rain. Plant the deeper areas with plants that thrive in soggy conditions, such as sedges and cattails.

With ten years of rain garden experience under his belt, Oregon landscaper Alfred Dinsdale has a different strategy for dealing with soils that drain very slowly. He notes that rain gardens filled with sandy soil are modeled on woodland bogs with anaerobic, acidic soils rich in organic matter. The problem is that the flow through a rain garden is pulsed, rather than slow and gradual, as it is in natural bogs. Because the sandy soil dries out between storms, the soil organisms that promote drainage in bog soils don't survive. The transition between clay and sandy soils can actually reduce infiltration rates over time.

In these cases, Dinsdale advises making your rain garden extra-large, but no deeper than 18 inches (45 cm). Loosen the soil at the bottom of the basin and mix in 3 inches (7.5 cm) of compost. If water sticks around longer than you'd like, lower the overflow and add another basin or just drain some water into the street.

As a last resort in very compacted soils, Dinsdale and other Portland landscapers build rain gardens over large basins filled with cobble or rubble. Known as dry wells, these basins fill quickly with water, which is stored in the spaces between the large rocks. Several layers of gravel gradually decreasing in size keep the rain garden soil from shifting down and filling the water-holding spaces between the rocks. Plants grow in a 9- to 12-inch (22.5- to 30-cm) layer of soil on top. South Carolina rain garden expert Sam Gilpin uses dry wells at some sites with clay soils. In the marshes around Charleston, the soil is mostly sandy with small lenses of clay in old floodplains. Gilpin sometimes digs dry wells through the clay layer into the sand layer, where water infiltrates quickly.

FINISHING TOUCHES: MULCH AND STEPPING STONES

Mulch is great for any garden. It suppresses weeds, shades and cools the soil, reduces evaporation, and encourages worms and other beneficial critters. For a rain garden, though, mulch is essential. Without it, the top layer of soil will dry out, forming a crust that repels water. Fungi and soil-dwelling creatures, which burrow through the soil and make pathways for rain to infiltrate, will shrivel and die in the harsh summer sun.

You can make mulch out of anything organic—cardboard, leaves, straw, coffee grounds—as well as rock. Most rain gardeners favor woodchips, coarse compost, or smooth river rock. Woodchips break down slowly and are often available free from tree trimmers, but they tend to float when a rain garden is flooded. Pecan hulls are a sustainable and beautiful alternative to cedar or bark mulch; use as a thin top layer over woodchips for a neater look. Compost is denser but often darker than woodchips and can get very hot in the sun. River rock obviously doesn't add organic matter to the soil, but it can add water. Dew condenses on the stones at night, and drips down into the soil. Rocks also heat up in the sun and increase the evaporation rate, making water in

the rain garden disappear faster. In the end, any mulch will work. Go with the look you like and what you can get cheaply or for free.

Spread mulch 3 to 6 inches (7.5 to 15 cm) thick. However, don't spread mulch right up to the base of fruit trees or perennials. Mulch can trap moisture near the trunk, causing crown rot in some species. If in doubt, leave a 1 foot (0.3 m) mulch-free space around woody perennials.

Also think about access. How will you get across your rain garden to the juicy tomatoes planted behind? How will you get in to weed? Do you want to create a small animal crossing? You can cross the rain garden on a pathway made of river rock, which lets water flow through. Or you could add large stepping stones, tall enough to be above water when the rain garden is full. Large stones can make for interesting visual accents as well.

RAIN GARDEN CONSTRUCTION CHECKLIST

- [ ] The bottom of the basin is level.
- [ ] The sides slope at about 60 percent.
- [ ] Holes were poked in the sides and bottom with a digging fork to improve infiltration before amending soil.
- [ ] The berm is level.
- [ ] The outlet is lower than the inlet and other points on the berm.
- [ ] The inlet and outlet are protected from erosion with rock or similar materials.
- [ ] Water flows through the conveyance system and into the rain garden. (Check with a hose if no rain is expected.)
- [ ] The entire basin and berm have been amended with compost and then mulched.

LAST WORDS

RESHAPING YOUR landscape to harvest rain is no small effort, but it will pay off many times over once your rain garden comes to life. In the next chapter, we explain how to choose and plant the grasses, shrubs, emergents, and flowers that will transform your mulched basins into beautiful, low-maintenance garden beds. The plant roots will penetrate the soil, host beneficial fungi, and, once they die and decompose, create compost-filled channels for water. Worms and insects will burrow into the rich, moist soil beneath the basin, creating more crevices where water will seep. As the seasons change and the plants mature, notice how often the basin fills and overflows. You may find the changes are dramatic.

PLANTING YOUR RAIN GARDEN

BY THE TIME YOU'RE READY TO

PLANT YOUR RAIN GARDEN, YOU'VE become intimately familiar with the contours of the land around your home, measured your roof and your rainfall, and flexed your digging muscles. For all gardeners, planting is both familiar and satisfying. For those planting a rain garden, the transformation from mulched depression to blooming oasis is dramatic. Back in the planning phase, you started with a vision of a natural habitat—forest, meadow, riverbank, or prairie pothole—that you wanted your rain garden to mimic. You may have a kept a running list of plants that thrive in local rain gardens or that you observed in natural depressions on local walks. If you haven't already, browse a local field guide with an eye for plants from your model habitat and make a list of plants that could work in your rain garden, aiming for a mix of water-loving emergents, grasses, and a few shrubs.

In this chapter we'll survey the many emergents, grasses, forbs, shrubs, and trees that make up a rain gardener's plant palette and discuss where in the garden to plant them. We'll also mention some plants that are particularly good for rain gardens in various climates, and talk about how and when to plant your rain garden. We mention species and cultivars that have done well in particular areas, but other species of the same genus are probably good candidates as well. Consult your local nursery for recommendations.

A rain garden has three planting zones: the center, the sloped sides, and the bermed edges.

▶ **The center is the wettest area of the rain garden. This planting zone may flood for up to 48 hours and will stay moist longer than any of the other zones. You can think of this zone as a seasonal wetland. Plants in the center of the rain garden must tolerate flooding, but also drought conditions between storms, so true marsh plants usually don't survive. Emergents, ferns, and water-loving native shrubs work well in the center.**

▶ **The sloped sides flood briefly, usually for less than 24 hours. For this zone, choose plants that tolerate damp soil but don't need much water during the dry season, such as deciduous shrubs, ferns, and grasses.**

▶ **Bermed areas and rain garden edges should never flood. This is the driest and best-drained zone, but the roots of plants in this zone can still tap into the moist soil retained in the sloped sides. This is why riparian plants— those that grow along a river's edge—work well in the berm. Any drought-tolerant perennials or self-seeding annuals can grow on the berm and around the edge.**

PREVIOUS SPREAD: A log serves as a bridge for small animals when the swale is full and a cool refuge for salamanders.

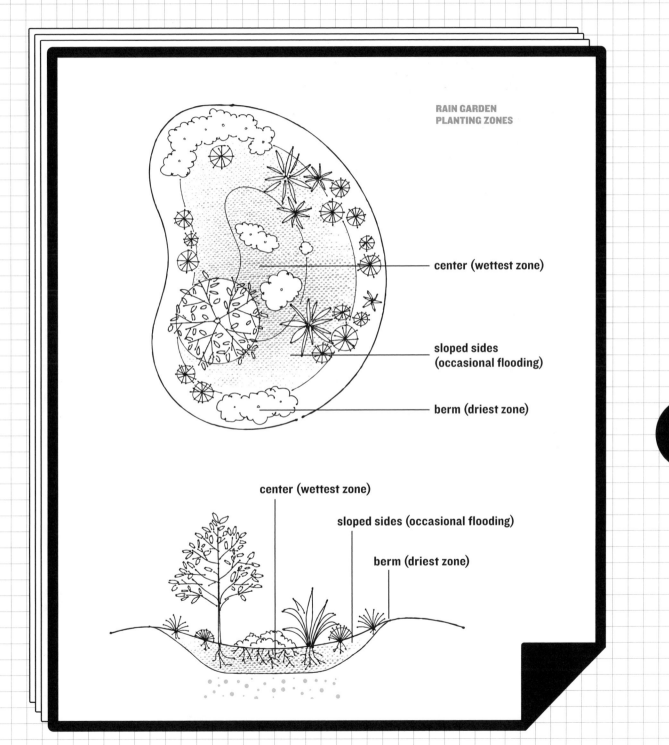

RAIN GARDEN
PLANTING ZONES

center (wettest zone)

sloped sides
(occasional flooding)

berm (driest zone)

center (wettest zone)

sloped sides (occasional flooding)

berm (driest zone)

119

COMMON RAIN GARDEN EMERGENTS

▶ **Tall sedge** (*Carex appressa*)

▶ **Bottlebrush sedge** (*Carex comosa*)

▶ **Slough sedge** (*Carex obnupta*)

▶ **Tussock sedge** (*Carex stricta*)

▶ **Spike rush** (*Eleocharis palustris*)

▶ **Common cottongrass** (*Eriophorum angustifolium*)

▶ **Knobby club-rush** (*Ficinia nodosa*)

▶ **Land quillwort** (*Isoetes histrix*)

▶ **Tapertip rush** (*Juncus acuminatus*)

▶ **Corkscrew rush** (*Juncus effusus*)

▶ **Yellow rush** (*Juncus flavidus*)

▶ **Torrey's rush** (*Juncus torreyi*)

▶ **Green bulrush** (*Scirpus atrovirens*)

▶ **Small-fruited bulrush** (*Scirpus microcarpus*)

▶ **Soft-stem bulrush** (*Scirpus validus*)

▶ **Dwarf cattail** (*Typha minima*)

EMERGENTS

So-called because they emerge above the surface of a pond or stream, emergents like full or partial sun and are ubiquitous in marshy and seasonally wet areas. Consult a field guide to wild plants for the species that are native to your area.

Many species of *Typha* (commonly known as bulrush in the United Kingdom and cattail elsewhere) can be found in temperate areas throughout the Northern Hemisphere, as well as parts of South America and Asia. They are good candidates for slow-draining rain gardens that receive regular rain, because they transpire a lot of water, but they will wither and die in rain gardens that experience extended dry periods. Be careful, though, because *Typha* can grow to 15 feet (4.5 m) and will take over without careful maintenance and pruning. For small gardens, consider a dwarf variety.

Sedges are usually bright green and grasslike, although the orange New Zealand sedge (*Carex testacea*) is bright orange. Sedges are very cold hardy: They will stay green down to 39°F (4°C).

Rushes are round in cross section and generally less than 2 feet (60 cm) tall. Notable ornamental varieties include the corkscrew rush (*Juncus effusus*), which grows in a spiral and can reach 3 feet (90 cm) in very wet areas. Bulrushes (*Scirpus*) grow to about 4 feet (120 cm) and have neat spiky seedheads. The zebra bulrush (*Scirpus lacustris* ssp. *tabernaemontani* 'Zebrinus') has eye-catching green and white striped leaves.

All emergents tolerate standing water and will do best in the wettest area of the rain garden, the center. Some rushes (such as *Juncus effusus* and *Juncus acuminatus*) can be planted in dry areas of the rain garden. Emergents grow in a spreading form, so a few plants go a long way. Rushes and sedges are more drought tolerant than cattails and bulrushes,

CASE STUDY

AN EDIBLE RAIN GARDEN WITH STEPPING STONES

by Dave Barmon

PLACE: Portland, Oregon, U.S.A.

HOMEOWNERS: Barmon Family

DESIGNERS: David Barmon, Fiddlehead LLC

INSTALLED: Spring 2008 and spring

CATCHMENT AREA: Front and backyard each 480 square feet (45 square meters)

RAIN GARDEN SIZE: 80 square feet (7.5 square meters) and 54 square feet (5 square meters)

ANNUAL RAINFALL: 42 inches (105 cm)

Local Columbia River basalt boulders double as stepping stones and structural design elements among the columnar aspen trees in this Portland, Oregon, rain garden.

A rain garden can be a focal point in the landscape, as well as a useful stormwater treatment system that hydrates the soil and plants throughout the seasons. Because I'm a landscaper, I wanted my home rain garden to showcase rain gardens' aesthetic and useful potential. Each rain garden receives water from half my roof and is sized to infiltrate water from a 2-inch (5-cm) storm event.

I installed the front rain garden first. I piped the

CONTINUED

which will turn brown during dry periods and then resprout in the rainy season. Plant sedges and rushes along the slopes and near the inflow—they will thrive in the moist soil and prevent erosion.

ORNAMENTAL GRASSES

Bunchgrasses are perennial grasses that tend to grow in discrete tufts or clumps rather than in sod-like carpets. They thrive in sunny rain gardens, and many are quite showy. Their fibrous roots grow up to 20 feet (6 m) deep in search of water and greatly improve infiltration. Because bunchgrasses are perennial, each season their roots push deeper into the soil. When the roots die off and decompose, they leave small, compost-filled channels that facilitate air and water flow into the soil. After several growing seasons, soil texture and organic matter improve dramatically, so that even compacted and clayey soils can support a broad array of plants.

If your rain garden mimics a prairie or meadow, grasses may dominate the moderate and dry zones (the slopes and berm). North America, Australia, and New Zealand hosts hundreds of bunchgrass (also known as tussock grass) species, but many native varieties are endangered because agriculture and development has replaced prairie and savannah ecosystems. Planting native grasses is a great way to preserve endangered varieties; contact your local native plant society for locally adapted seed or seedlings.

Most grasses need full sun, but some grow in partial or full shade, among them river oats (*Chasmanthium latifolium*), which has pretty, dangling seedheads, and moor grass (*Molinia caerulea*), which has variegated green and white leaves. Big bluestem (*Andropogon gerardii*), native to the Great Plains, grows to 8 feet

COMMON RAIN GARDEN GRASSES

- Big bluestem (*Andropogon gerardii*)
- Purple three-awn (*Aristida purpurea*)
- Meadow pinegrass, reedgrass (*Calamagrostis canadensis*)
- River oats (*Chasmanthium latifolium*)
- Tufted hairgrass (*Deschampsia caespitosa*)
- Meadow barley (*Hordeum secalinum*)
- Basket grass (*Lomandra longifolia*)
- Japanese silver grass (*Miscanthus sinensis*)
- Moor grass (*Molinia caerulea*)
- Gulf muhly grass (*Muhlenbergia capillaris*)
- Switchgrass (*Panicum virgatum*)
- Fountain grass (*Pennisetum alopecuroides*)
- Esparto grass (*Stipa tenacissima*)

water 8 feet (2.4 m) away from the house in a drainpipe, which spills into a small depression next to a large boulder. Most of the time, this small basin holds all the runoff. In heavy rain, overflow from the small basin runs through a small swale and into the main rain garden. I chose many edible plants for this space: lingonberry (*Vaccinium vitis-idaea*), cranberry (*Vaccinium macrocarpon*), and camas (*Camassia*) are planted in the bottom of the rain garden, under a medicinal cascara tree (*Rhamnus purshiana*). I amended the soil with arborvitae needles and peat moss. To give the water-loving lingonberries a boost during the dry Oregon summer, I buried slabs of cedar (cut with a friend's portable sawmill) in the ground. The decomposing wood will act as a sponge. I planted ostrich fern (*Matteuccia struthiopteris*), blueberry (*Vaccinium ovatum*), and wintergreen (*Gaultheria*) on the sloped sides and a Fuyu persimmon tree (*Diospyros kaki* var. *fuyu*) on the berm.

For the back rain garden, I first created a design that was a rough sketch, but more or less to scale, that showed hardscape and trees. Next, I excavated the soil 6 inches (15 cm) below the final depth I wanted—about 18 inches (45 cm) in total. I did not do any infiltration

tests because I am already familiar with the earth in my neighborhood; it's a prehistoric gravely floodplain that drains like a sieve. Any overflow from the basin runs down a gentle, vegetated slope.

After digging out the area to the correct grade, I placed several small Columbia River basalt boulders from a local quarry. The flat-topped boulders become stepping stones when the basin fills with water. In its natural form and in the hands of skilled stoneworkers, stone is an amazing and timeless material. Columbia River basalt is lava that cooled into gray columns. The nearest quarry is 30 minutes from my home. I often use basalt in rain gardens to make the space visually interesting and to build stepping stones and small check dams to slow the flow of water.

The south-facing backyard rain garden lies 10 feet (3 m) from my house and features five quaking aspens (*Populus tremuloides*). Aspens tolerate wet roots and grow in a columnar shape, and these will shade our house in the summer but let winter light in. I planted an understory of low-growing mop-headed sedge (*Carex caryophyllea* 'The Beatles'), nodding onion (*Allium cernuum*), purple coneflower (*Echinacea purpurea*), common camas (*Camassia*), sword fern

(*Polystichum munitum*), deer fern (*Blechnum spicant*), and salal (*Gaultheria shallon*).

I installed irrigation to help the plants get established, but plan to stop watering after the second year. Over the next few years, I want to install a greywater system and cistern to save even more water. In the meantime, my two rain gardens are keeping 24,000 gallons (90,850 L) of water out of the city storm pipes every year, which is a great start to my water-harvesting adventure.

On a personal note, I believe that we can all promote change in the world on different scales. Harvesting water from your roof and directing it into the soil is a positive step toward social, environmental, economic, and political change. This simple act doesn't require any intense convictions or political leaning. All humans can understand that water should go into the ground to help our plants grow.

COMMON RAIN GARDEN HERBACEOUS PERENNIALS

▶ **Yarrow** (*Achillea millefolium*)

▶ **Swamp milkweed**
(*Asclepias incarnata*)

▶ **New England aster**
(*Aster novae-angliae*)

▶ **Blue grass lily**
(*Caesia calliantha*)

▶ **Common everlasting**
(*Chrysocephalum apiculatum*)

▶ **Flax lily** (*Dianella*)

▶ **Western bleeding heart**
(*Dicentra formosa*)

▶ **Purple coneflower**
(*Echinacea purpurea*)

▶ **Yellow flag iris**
(*Iris pseudacorus*)

▶ **Lavender**
(*Lavandula angustifolia*)

▶ **Prairie blazing star**
(*Liatris pycnostachya*)

▶ **Great blue lobelia**
(*Lobelia siphilitica*)

▶ **Purple loosestrife**
(*Lythrum salicari*)

▶ **Bistort** (*Persicaria bistorta*)

▶ **Black-eyed Susan**
(*Rudbeckia hirta*)

▶ **Stiff goldenrod**
(*Solidago rigida*)

▶ **Grass triggerplant**
(*Stylidium graminifolium*)

▶ **Tufted bluebell**
(*Wahlenbergia communis*)

(2.4 m), resists drought, and turns tawny in autumn. Plant breeders have developed cultivars of switchgrass (*Panicum*), fountain grass (*Pennisetum alopecuroides*), and Japanese silver grass (*Miscanthus sinensis*) with different heights and bloom sequences.

All bunchgrasses are drought tolerant and prefer full sun. Plant them on the slope or berm of the rain garden, where they'll protect against soil erosion. (In arid regions, plant grasses in the wet zone at the bottom of the basin.) Use grasses as accents around the border of the rain garden to mimic a natural pond. Tufted hairgrass (*Deschampsia caespitosa*) and switchgrass (*Panicum*) tolerate flooding and can control erosion near the rain garden inflow.

HERBACEOUS PERENNIALS

Herbaceous perennials grow and flower season after season. During the winter, the tops die back but the roots stay alive underneath the soil or snow, awaiting the warmth of spring to bloom again. They bloom for a few weeks to a few months, so choose a mix of herbaceous perennials that will sustain bloom and color from spring to autumn. Herbaceous perennials are a good complement to shrubs, trees, and grasses, which add texture, variety, and color in seasons when the herbs don't bloom. Thousands of flowering perennial species exist in wide enough variety to find something suitable for any site. Herbaceous perennials need good drainage, so plant them on the berm or around the edge of the rain garden basin, unless you have very well drained soils.

OPPOSITE:
Ostrich fern
(*Matteuccia struthiopteris*)

COMMON RAIN GARDEN FERNS

- **Giant leather fern** (*Acrostichum danaeifolium*)
- **Southern maidenhair fern** (*Adiantum capillus-veneris*)
- **Hart's tongue fern** (*Asplenium scolopendrium*)
- **Lady fern** (*Athyrium filix-femina*)
- **Fishbone water-fern** (*Blechnum nudum*)
- **Moonwort** (*Botrychium lunaria*)
- **Male fern** (*Dryopteris filix-mas*)

- **Sensitive fern** (*Onoclea sensibilis*)
- **Cinnamon fern** (*Osmunda cinnamomea*)
- **Royal fern** (*Osmunda regalis*)
- **California polypody fern** (*Polypodium californicum*)
- **Sword fern** (*Polystichum munitum*)
- **Marsh fern** (*Thelypteris palustris*)
- **Virginia chain fern** (*Woodwardia virginica*)

FERNS

Because ferns reproduce with spores rather than seeds, they need very moist conditions during the sporulating season. Many ferns tolerate dry conditions at other times of the year, and most ferns grow well in the shade. Ferns range in size from the diminutive maidenhair (*Adiantum*), which likes moist soil and deep shade, to the towering Western sword fern (*Polystichum munitum*), which thrives in coniferous forests and can reach 6 feet (1.8 m) in height. Ferns native to your ecological region are most likely to survive through the dry season. If you want to experi-

ment with more moisture-loving species, plant them near the inflow of your rain garden. Plant ferns in any rain garden zone, but make sure they will be shaded for most of the day.

GROUNDCOVERS

Groundcovers are low-growing perennials, either woody or herbaceous, that grow in a low, spreading form. Some are low-growing varieties of evergreen shrubs, such as creeping Oregon grape (*Mahonia repens*) and dwarf madrone (*Arbutus unedo* 'Compacta'). Others spread by runners, such as wild strawberry

(*Fragaria*). Choose one or at most two groundcovers for your rain garden, and plant just two or three plants. They will soon spread to fill in the spaces between larger plants.

SHRUBS

Deciduous and evergreen shrubs become larger focal points in a garden and define points of visual interest. They provide structure and color—spring or autumn blooms, leaves that change color throughout the growing season, or colorful winter berries. When strategically planted as a hedge along part of the rain garden border, they can act as a windbreak or privacy screen. Windbreaks keep downwind plants from drying out and provide protected habitat that improves pollination by butterflies and bees. The spreading form of low-growing shrubs helps stabilize slopes and reduce erosion.

Deciduous shrubs that grow along streams or lakeshores are good candidates for rain garden basins and slopes. Any ornamental shrub will grow on the rain garden berm, but dogwood (*Cornus*), mock orange (*Philadelphus*), and blueberry (*Vaccinium ovatum*) will tolerate at

Common rain garden grasses, including blue oatgrass (*Helictotrichon sempervirens*), Mexican feather grass (*Nassella tenuissima*), and tufted hairgrass (*Deschampsia caespitosa*).

127

COMMON RAIN GARDEN GROUNDCOVERS

▶ **Sweetflag** (*Acorus calamus*)

▶ **Wild ginger** (*Asarum caudatum*)

▶ **Crossvine** (*Bignonia capreolata*)

▶ **Spring sedge** (*Carex caryophyllaea*)

▶ **Wild clematis** (*Clematis virginiana*)

▶ **Firewitch** (*Dianthus gratianopolitanus*)

▶ **Wild strawberry** (*Fragaria chiloensis*)

▶ **Hop goodenia** (*Goodenia ovata*)

▶ **Running postman** (*Kennedia prostrata*)

▶ **Cascade Oregon grape** (*Mahonia nervosa*)

▶ **Creeping boobialla** (*Myoporum parvifolium*)

▶ **Passionflower vine** (*Passiflora incarnata*)

▶ **Frogfruit** (*Phyla incisa*)

▶ **Blue moor grass** (*Sesleria caerulea*)

▶ **Creeping blueberry** (*Vaccinium crassifolium*)

least occasional flooding. Be sure to keep mature height in mind when choosing shrubs. They grow quickly, can smother or shade out other rain garden plants, and require extra work to prune or transplant.

Evergreen shrubs provide year-round green foliage but also produce flowers (such as rhododendrons and azaleas, *Rhododendron*), or attractive berries (beauty berry, *Callicarpa*). Many evergreen shrubs require well-drained soils, but some, such as arborvitae (*Thuja occidentalis*), can tolerate wetter soils.

In all but the largest rain gardens, plant no more than one full-sized shrub or tree, whether evergreen or deciduous. Plant breeders have developed dwarf varieties of many shrubs, but dwarf is relative. According to the American Conifer Society, for example, a dwarf can grow 6 inches (15 cm) per year and be as tall as 6 feet (1.8 m) after ten years. Always confirm mature height before planting.

COMMON RAIN GARDEN SHRUBS

- **Kangaroo paw** (*Anigozanthos*)
- **Dogwood** (*Cornus*)
- **Digger's speedwell** (*Derwentia perfoliata*)
- **Oceanspray** (*Holodiscus discolor*)
- **Holly** (*Ilex*)
- **Virginia sweetspire** (*Itea virginica*)
- **Cushion bush** (*Leucophyta brownii*)
- **Spicebush** (*Lindera benzoin*)
- **Wax myrtle** (*Myrica*)
- **Common riceflower** (*Pimelea humilis*)
- **Azalea** (*Rhododendron*)
- **Rhododendron** (*Rhododendron*)
- **Sumac** (*Rhus*)
- **Currant** (*Ribes*)
- **Thimbleberry** (*Rubus parviflorus*)
- **Lilac** (*Syringa vulgaris*)

TREES

Specimen trees are particularly striking for their form, blossoms, or color. One such tree that can grow anywhere in a rain garden is Sitka dwarf alder (*Alnus viridis* ssp. *sinuata*), a slender, graceful tree with striking dark bark that grows to 15 feet (4.5 m). Japanese maple (*Acer palmatum*), with its twisting branches and colorful autumn foliage, will thrive on a rain garden berm. Plant vine maple (*Acer circinatum*), native to western North America, anywhere in the rain garden. Its leaves turn fire engine red in open areas or yellow in partial shade. Mountain ash (*Sorbus*) grows to 13 feet (4 m) and has clusters of bright red berries that stay on the tree all winter and that birds love. Plant it on the berm or slopes. Willows (*Salix*) deserve special note. They thrive in rain gardens, and there are hundreds of species in different shapes and sizes to choose from. Some willows can grow quite

OPPOSITE: Wild strawberry (*Fragaria*), an effective groundcover.

large, but all willows can be coppiced. Coppicing involves cutting the tree back to a single trunk or stump, then letting the branches resprout. Willow branches are great for basketry or for building trellis, garden furniture, or play structures for kids.

You can plant water-loving trees as an overstory for a shady rain garden, or, in some cases, make your rain garden under an existing tree. Most trees will tolerate the extra water as long as the soil around some roots stays dry. Note, however, that cherries and their wild relatives exude a root toxin when flooded that can kill other plants, and avocado trees will die if their roots stay wet.

PLANTING DESIGN

Rain garden design involves balancing visual aesthetics with plants' water needs and habitat value. Aesthetics are a matter of personal preference. Choose colors and textures that appeal to you, or consult one of the hundreds of garden design books for inspiration. Make sure your plant list includes plants that tolerate wet and dry conditions and features a mix of emergents, groundcovers, grasses, and herbaceous perennials, with a fern, shrub, or tree thrown in if you like.

Colorful flowers and foliage brighten up gardens and change from season to season. Some gardeners choose one or two colors to highlight, then select deciduous shrubs and flowering perennials based on this selection. In general, colors opposite each other on a traditional color wheel work well together, such as yellow and purple or orange and blue. White flowers and variegated foliage stand out and are effective in brightening up shady spaces. Gardens that incorporate lots of white are sometime considered moon gardens for their ability to glow in the moonlight.

When water spills from basin to basin or runs down channels lined with smooth river rock, the

COMMON RAIN GARDEN TREES

▶ **Vine maple** (*Acer circinatum*)

▶ **Japanese maple** (*Acer palmatum*)

▶ **Sitka dwarf alder**
   (*Alnus viridis* ssp. *sinuata*)

▶ **River birch** (*Betula nigra*)

▶ **White ash** (*Fraxinus americana*)

▶ **Swamp paperbark** (*Melaleuca ericifolia*)

▶ **Desert ironwood** (*Olneya tesota*)

▶ **Pacific ninebark**
   (*Physocarpus capitatus*)

▶ **Black poplar** (*Populus nigra*)

▶ **Velvet mesquite** (*Prosopis velutina*)

▶ **Willow** (*Salix*)

▶ **Mountain ash** (*Sorbus*)

131

OPPOSITE:
Two rainwater cascades and several contour swales are hidden in this dry-stacked stone retaining wall.

HOW TO AVOID DROWNING YOUNG PLANTS

RAIN GARDEN plants need good drainage until they establish a strong root system, at which point they can tolerate extended flooding. In slow-draining soils, you may want to lower the outlet for the first rainy season, thereby lowering the pooling depth and reducing the time that plants will be flooded. Once plants are growing vigorously, raise the outlet to its design height.

rain garden comes alive. As you design, think about how plants can accent flowing water or still pools. Consider planting one larger shrub or tree at the edge of the basin. Willow, dogwood, rhododendron, azalea, alder, ninebark, Japanese maple, or weeping mulberry are all good candidates for this specimen tree. In shady rain gardens, tuck ferns in at the base of large rocks and plant mosses in the crevices between rocks—in moist climates, they will spread to cover the rocks. In sunny rain gardens, use large, clumping sedges and irises to highlight large stones.

Weeds may be more of a problem in rocky areas of the rain garden. Don't use a weed barrier to suppress them, though, because it will eventually clog with sediment and reduce infiltration into the soil below. Instead, while building the garden lay down several layers of cardboard on top of the soil, wet the cardboard, and place a 4-inch (10-cm) layer of woody mulch under the top rock layer. If your design uses many rocks, dig in compost around individual plants rather than spreading it evenly across the basin.

As you arrange your plants, first consider their ultimate size and shape as well as color. Landscape architects think in layers. The idea of layers incorporates plants' relative sizes as seen from a vantage point. For example, if your rain garden runs along a fence, place large shrubs along the fence, tall grasses and bushy perennials in front of the shrubs, and groundcovers in the front. If you plant a tree in your rain garden (or place your rain garden near an existing tree) consider planting large shrubs near the tree and shorter plants farther away.

Make sure each plant will get the right amount of water. In moist climates, plant plenty of emergents in the wet zone on the basin floor. In dry climates, plant grasses and water-loving herbaceous perennials in the wet zone. Many shrubs will also grow in the wet zone, while tolerating dry spells. Willows are an exception: They need near-constant moisture, so don't plant them in very well drained soils or in rain gardens with an under drain.

OPPOSITE:
Rain garden filled with herbaceous perennial.

133

Ornamental grasses and herbaceous perennials do best on rain garden slopes and berms. Most plants can grow on the berm or near the rain garden, but not all. Avocadoes are very sensitive to waterlogging and may get root rot in the moist soil near a rain garden. Established citrus and dryland oak trees can suffer if they suddenly get more water, although they usually survive as long as some roots stay dry. Cherry trees can grow near a rain garden, but not in the basin, because when their roots get flooded they release a toxin that will kill the tree.

Finally, consider wildlife when choosing plants for your rain garden. Most songbirds and small mammals prefer the dense cover provided by shrubs. They may eat insects; seeds of grasses, flowers, and emergents; or fruits of groundcovers, trees, and shrubs. Snakes and lizards like sunny rocks to bask on, and amphibians hide in the moist soil under shaded logs. If you want to attract wildlife to your garden, plant species that provide food and shelter for particular birds, mammals, amphibians, reptiles, or insects. Then arrange these plants to provide dense cover, open areas, and sunny and shady spots.

WHEN AND HOW TO PLANT YOUR RAIN GARDEN

The best time to plant your rain garden is during a cool season with regular rain. If you plant in the heat of summer, be prepared to water frequently during this initial growing season to help plants cope with heat stress. Watch how your rain garden drains after the first big storms, and temporarily lower your outlet if the plants seem to be getting too much water.

The best time of the year to plant a rain garden varies among different climates:

ABOVE:
Slough sedge (*Carex obnupta*) and Douglas spirea (*Spirea douglasii*) grows in swampy areas and thrives in slow-draining rain gardens.

OPPOSITE:
Children carefully place a groundcover seedling into a rain garden.

▶ In temperate, cold-winter climates, plant in early autumn at least one month before the first hard frost or in the spring. This will give plants a chance to grow new roots before the ground freezes. Add a thick layer of mulch around plants to insulate roots from the cold air.

▶ In tropical and subtropical climates, plant your rain garden after the end of hurricane season or in winter or early spring. Tropical storms bring heavy rains that can saturate soils and cause tender roots to rot.

▶ In dry-summer climates, plant after the first autumn rains or in spring before the rains have ended. If plants seem to be getting too much water, lower the outflow for the first season.

The best part about planting a garden is that anyone can do it. Involve kids and elders, and invite gardener friends over. Planting a rain garden is no different than planting a perennial bed. Dig a small hole, gently remove the plant from its pot, and loosen any bound roots. Place the plant in the hole, replace the soil while being sure not to cover the root crown, and firmly press around the base of the plant to push out any trapped air. Then water well, unless rain is on the way.

Here are some planting tips:

▶ When you plant shrubs and trees, be careful not to bury the trunk flare (the swollen part of the stem or trunk just above the roots). Wet soil around the trunk flare can cause rot, which can kill the plant.

▶ You can plant the berm and upper slope of the rain garden from seed, but seedlings work best in the basin. Note, however, that seeds tend to float and scatter if rains arrive before the seedlings take root.

▶ As you plant, try to avoid compacting the soil. This can be difficult during a large work party. Install temporary stepping stones made of scrap plywood to spread your weight, and loosen the soil with a digging fork once you're done planting.

▶ Some rain gardeners prefer to mulch before planting, while others spread mulch around newly planted plants. Whatever your preference, make sure you give your rain garden a 3-inch (7.5-cm) layer of mulch.

▶ Any mulch will work in your rain garden, but woodchips and leaves will float when the basin fills, and settle back onto small seedlings when the water drains away. Many rain gardeners recommend using a denser mulch, such as pecan hulls or woody composts. Once plants are established, any kind of mulch is suitable.

▶ Water new seedlings well with a hose, and water as necessary until the plants are established.

"

LAST WORDS

ONCE PLANTS get established, they will survive on rainfall alone, but young seedlings will need supplemental watering in most climates. If plants start to wilt, they may need more water. But if you've had lots of rain and the soil feels moist, they may be suffering from too much water. If plants die, check for root rot or mold or slime around the roots—sure signs of overwatering. If too much rain is to blame, lower the rain garden outlet until plants are growing vigorously. At that point, they should be able to tolerate flooding for several days, and you can raise the outlets to its design height. In the next chapter, we'll walk you through typical garden maintenance and describe what to do if problems arise.

PLANTING WORKSHEET

Before you select plants for your rain garden, consider the information you gathered during site assessment and design.

I. On a piece of graph paper, sketch the basin shape and play with placement. Mark the inlet and the outlet, and outline the area within the garden that is lower than the outlet (that is, the sloped sides and the center).

2. Shade in and label the areas that receive full or partial shade.

3. Copy the following information from your site assessment and design worksheets:

Rain garden goals or dreams (edibility, color and texture, wildlife habitat): _____

Rain garden area (square feet or square meters): _____

Maximum pooling depth (inches or cm): _____

Estimated average time to drain (in hours or days): _____

4. Collect your plant lists, and in the table note their sun and drainage requirements, mature height, and color.

PLANT NAME	PLANT TYPE*	SUN AND SHADE NEEDS	SOIL AND DRAINAGE NEEDS	MATURE HEIGHT	NOTES (COLOR, EDIBLE PART)

* Emergent grass, herbaceous perennial, fern, groundcover, shrub, or tree.

5. Call local nurseries and visit native plant sales to see which plants are available.

6. From the final list of plants, divide plants by rain garden zone (berm, slope, or center).

7. Arrange plants in the basin. You can use paper circles cut to different sizes to represent plants. Place plants where they will get the amount of sun and water that they need. Remember to allow room for the plants to grow to their full size.

MAINTAINING YOUR RAIN GARDEN

RAIN GARDENS REQUIRE

LESS MAINTENANCE THAN LAWNS, vegetable gardens, or ornamental beds. You'll spend most time adding mulch and perhaps pruning if you prefer a manicured look. You'll also need to keep an eye on the infrastructure of your rain garden—the gutters, downspout, swales or buried pipe, inlet, and outlet—to make sure nothing is clogged, disconnected, or eroding. As you tend your rain garden, tread lightly to avoid compacting the soil. Work from the edges and outside the berm or from stepping stones.

Fertilizers are unnecessary in rain gardens and will only encourage weeds. Native plants are adapted to local soils and climate, and most rain garden cultivars are well suited to moderate levels of nutrients. Adding fertilizers will encourage non-natives without helping native plants and add nutrients to runoff that does leave the rain garden.

FIRST-YEAR MAINTENANCE

PREVIOUS SPREAD: Rain gardens host beneficial insects like this bumble bee.

Your rain garden will require the most maintenance during the first year, and devoting time to these tasks now will make maintenance much easier in the long run. The tasks include weeding, watering plants during dry spells, and keeping an eye out for erosion, clogging, or poor flow in the conveyance system. It's especially important to watch how your rain garden fills and drains during the first few storms. Most rain gardens require some small adjustments at first. Don't be afraid to raise or lower the outlet or modify the basin if it seems too small or large for the rain it receives. If plants seem to be suffering from too much or too little water, transplant them to a different location.

WEEDING

Weeding is the main task during a rain garden's first year or two. Mature rain gardens need little or no weeding, but young plants can't compete with weeds. Before young rain garden plants and natives have a chance to grow deep roots and fill in the space, weeds will try to take over bare spots. Pull weeds out by the roots before they drop seed. Keep your rain garden mulched with at least 3 inches (7.5 cm) of woodchips or coarse compost, which will suppress weeds and make the ones that do sprout easier to pull out. Your hard work early on will pay off for years to come.

WATERING

Regular watering during the first year will help young plants establish healthy root systems.

Once the plants in your rain garden get established, you shouldn't have to water. During the first year, check plants and soil weekly for wilting or other signs of water stress.

CHECKING THE OUTLET AND OVERFLOW DRAIN

You designed the overflow drain for your rain garden to maximize rain infiltration during large storms. Once rain garden plants are mature, they can tolerate standing water for several days, but extended flooding can damage young seedlings and transplants. If you grow plants from seed or if you see plants suffering from too much water, lower the bottom of the outlet 1 inch (2.5 cm) or more to reduce pooling depth. After the first year (or possibly two if you start plants from seed), raise the bottom of the outlet to its original design height.

SEASONAL TO ANNUAL MAINTENANCE

Depending on your climate, you may experience two, three, four, or more seasons. Seasonal maintenance includes tasks that may need doing more than once a year—usually just before and after periods of heavy rain. Inspect the gutters, downspouts, pipes, inlet, and outlet for clogging before the rainy season to ensure that rain can flow freely to the rain garden. Perform most garden-related maintenance after the end of the rainy season, when plants and weeds are growing vigorously and heavy rains may have washed mulch or soil away.

CHECKING THE CONVEYANCE SYSTEM

Inspect your gutters and downspouts for debris buildup or clogging. If leaves have clogged the gutters or downspouts, consider installing screens on your gutters. Observe water flow through downspouts, pipes and swales during a rainstorm, or use a hose to make sure water can still run freely. Pipes should empty onto a hard surface (such as a large flat stone or cobble splashguard) to reduce erosion. Channels and swales should be cleared of weeds and other debris.

CHECKING THE INLET AND OUTLET

Take a moment before and after each rainy season to inspect your rain garden inlets and outlets. Look for a path for clear access to water flow. If you see any sediment buildup or signs of erosion, be sure to address these right away. Remove any sediment that has washed down into the rain garden so that water can flow in and out freely. If you notice erosion, take time to fill in the eroded spots with rocks to reduce further erosion.

CHECKING THE BERM

Check the edges of your rain garden, especially any bermed areas, for signs of erosion or settling. Add material to bring the berm back to its original height. Try to determine the cause of any erosion, and reinforce eroding areas with cobble or river rock.

ADDING MULCH

Mulch keeps the soil moist, which helps water infiltrate faster. Unmulched rain gardens develop a hardpan surface when hot sun dries the bare soil surface between rains, and hardpan drastically reduces infiltration. Mulch also keeps plant roots cool, reduces erosion, and suppresses weeds.

Once a year, add 2 to 3 inches (5 to 7.5 cm) of woody compost around plants. Compost helps the soil retain moisture and supplies nutrients to rain garden plants. Be careful when using lightweight bark mulch below the rain garden outlet. It will float around when the rain garden fills up, which makes a mess and can smother new plants. Bark or woodchips are fine for mulching the berm.

141

WEEDING

After the first year, expect to weed only occasionally. As your rain garden plants grow deep roots, fewer weeds should appear. Once the rain garden fills in, the mature plants will shade out weeds. If you notice weeds or bare soil, check to see whether water flowing into or out of the rain garden is causing erosion. If so, add river rock, mulch, or more plants to protect the soil from moving water.

PRUNING

How much you prune depends on your gardening style. Gardeners who enjoy a more formal look usually enjoy keeping their rain garden tidy by shaping and deadheading in the autumn or winter. Other gardeners may prefer a wilder look and choose to forgo pruning and let plants take on bushy or rangy forms. Pruning will not affect the rain garden's infiltration function, but it will affect its habitat value. Songbirds and other small wildlife like to hide in bushy thickets, so thinning will make your rain garden less attractive to some species.

CUTTING BACK AND DEADHEADING PLANTS

Leave dead stems and seed heads in place through the winter to provide food and shelter for local wildlife. When new growth emerges in spring, cut back the past season's dead growth to the ground. Cut up the dry material you pruned and add it to the woody mulch in your rain garden, or compost it elsewhere.

DIVIDING PLANTS

As new growth comes in, you will most likely see that herbaceous perennials have spread. Even after the first year, you may find it necessary to divide some plants and pass them onto friends. This is a great time to encourage a friend or neighbor to put in a rain garden. What better way to start than with a knowledgeable friend and free plants?

SIGNS THAT YOUR RAIN GARDEN NEEDS MAINTENANCE

Prolonged standing water (that is, for more than 24 hours) requires immediate attention. For clay soils, lower the overflow to reduce pooling depth or expand the size of the basin. For loamy soils, prepare a mix of 60 percent sand and 40 percent compost and till this into the upper 3 to 6 inches (7.5 to 15 cm) of the rain garden. Make sure to top with a 3-inch (7.5-cm) layer of mulch.

If you see plants dying at any time of year, note the species and its location in the rain garden, and try to determine what caused its demise. Are other specimens of the same plant thriving? If so, where in the rain garden are they growing? Try replanting the same species or cultivar in a different zone of the rain garden. If the plant seemed to die from too much water or root rot, move it to a drier zone. If the plant seemed to dry out, wilt, or turn brown, try transplanting to the basin floor. Like all gardening, rain gardening requires trial and error. If repeated trials of a particular plant fail, try a different species.

LAST WORDS

AFTER THE FIRST year, your rain garden should need little care. If, like many rain gardeners we've met, you find yourself looking for more ways to harvest rain in your home landscape, read on. In the next chapter, we present lots of ways to accessorize your rain garden—with living roofs, flagstone patios, rain barrels, and more. We also urge you to look beyond your roof to the street and sidewalk along the street and harvest the rain falling there.

WORKSHEET

Annual Maintenance Checklist

☐ . . . . CLEAN GUTTERS AND DOWNSPOUTS OF DEBRIS.

☐ . . . . OBSERVE WATER FLOW THROUGH DOWNSPOUTS, PIPES,
AND SWALES AND CLEAR OBSTRUCTIONS.

☐ . . . . REPAIR ANY EROSION AND REMOVE WEEDS IN DIVERSION SWALES.

☐ . . . . INSPECT RAIN GARDEN INLETS AND OUTLETS FOR SIGNS
OF SEDIMENT BUILDUP OR EROSION.

☐ . . . . CHECK THE BERM FOR SIGNS OF SETTLING OR EROSION;
ADD SOIL IF NECESSARY.

☐ . . . . ADD COMPOST OR OTHER TYPES OF MULCH YEARLY.

☐ . . . . CHECK MATURE RAIN GARDEN FOR WEEDS AND BARE SOIL,
WHICH MAY BE SIGNS OF EROSION.

☐ . . . . PRUNE, DEADHEAD, AND DIVIDE PLANTS AS DESIRED.

YOUR RAIN GARDEN AS PART OF AN INTEGRATIVE DESIGN

145

IF YOU'VE WORKED THROUGH

THE BOOK THIS FAR, YOU HAVE A RAIN garden that is capturing runoff, recharging the aquifer, and restoring a bit of native habitat in your landscape. Like many rain gardeners, you might be thinking of more ways to use the rain and to keep pollutants from entering local waterways. Maybe you're eyeing more catchment areas—the patio, the driveway, even your neighbors' downspouts—or wondering how to use rainwater to irrigate fruit trees or vegetables. In this chapter, you'll learn how to catch every drop of rain that falls on your home watershed: in basins and swales, on living roofs and walls, in cisterns, and in curbside planters that harvest street runoff. We'll also look at water-wise gardening techniques and describe how to tap the rain and your dirty wash water (known as greywater) for irrigation before turning to the hose. We'll tie all of these strategies together with the tool of integrative design.

Your roof is probably the largest source of runoff on your property, followed by a driveway or patio. Of course, natural surfaces shed water as well. Rocky outcrops, boulders, and hardpan soils shed virtually all of the rain that falls there, and bare soil and lawn shed quite a bit, too. The trick to catching every drop of that rain is to use or store it in many dispersed locations—on living roofs, in dry wells under permeable patios, and in cisterns—and then infiltrate the rest into rain gardens. By spreading rain throughout your home watershed, you turn the landscape into a giant sponge that can water and fertilize a productive and sustainable home landscape.

AN INTEGRATIVE DESIGN PRIMER

Integrative design starts with thoughtful observation. This means lounging in your backyard, relaxing, and noticing how the surrounding ecosystem is functioning on its own. First, consider every living and nonliving element—what it consumes, what it produces, and what it inadvertently provides. Then think about how you could arrange those elements to maximize the productive interactions between them.

In the broadest view, an integrative design considers all the flows of energy through a landscape—sun, wind, water, hot and cool air flows, plants, animals, noise, even airborne

PREVIOUS SPREAD:
Living roof on an outdoor kitchen.

INTEGRATED HOME
WATERSCAPE

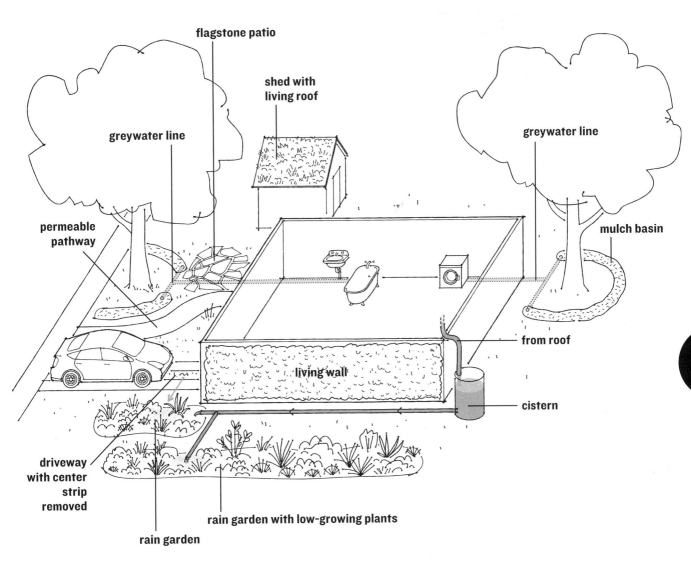

flagstone patio

shed with
living roof

greywater line

greywater line

permeable
pathway

mulch basin

from roof

living wall

147

cistern

driveway
with center
strip
removed

rain garden with low-growing plants

rain garden

THE ORIGINS OF INTEGRATIVE DESIGN

Our approach to integrative design is adapted from permaculture, a system developed by Australians Bill Mollison and Dave Holmgren in the late 1970s. Mollison and Holmgren noticed similarities in traditional agriculture, building, and watershed management practices worldwide:

▶ a basis in long and thoughtful observation of natural systems;

▶ an adaptive, constantly evolving approach to working on land; and

▶ collaborative relationships with plants, animals, and forces of nature.

Permaculture considers human settlement at all scales, from the home garden to the watershed scale, with a focus on agricultural systems that produce food, fuel, fiber, and medicine. In this book, we narrow the focus to design for water, because water supports plants and soil—the building blocks of a backyard ecosystem.

particles of dust and soot. The designer arranges buildings, fences, water tanks, pavement, paths, plants, earthworks, and ponds to create a beautiful, self-sustaining ecosystem. Like natural ecosystems, these created ecosystems provide many benefits to humans, including clean water stored in biomass and in the soil; cool, moist air transpired from plants; winter sun and summer shade indoors and out; and food, medicine, and wildlife.

The practice of integrative design is simple. In fact, you've already gone through each step as you designed your rain garden. First, carefully observe your home landscape throughout the year in all kinds of weather. Record the plants and animals you see as well as descriptions of the soil. Draw maps of sun and shade, slope, wind, water, and any other patterns you recognize. Take time to relax in your garden. Take naps there.

Next, think about the ways you want your home environment to function. Possibilities include passive summer cooling and winter heating; water for irrigation or indoor use; electricity from solar or wind power; shelter from street noise and pollution; gardens of food, medicine, craft materials, and wildlife habitat; and places to lounge, play, work, or cook outside.

List elements you could bring into your landscape. In terms of water, consider incorporating living roofs, permeable pavement, cisterns, and greywater systems. List the many functions each element offers and the connections each has with other elements. Then try to design redundancy into your home ecosystem. Natural ecosystems are redundant; for example, leaves, mulch, and fungal mycelia all intercept falling rain. In home landscape design, several natural features can combine to cool your house. You can plant deciduous trees or vines to shade south- and west-facing windows during the summer, open windows to catch breezes, paint your house a light color to reflect sun, and install a living roof for evaporative cooling and insulation.

Next, arrange elements until each plant,

Sedums thrive in harsh, dry rooftop conditions, resist fire, and stay green year round.

water source, and structure is performing as many functions as you can think of. The more functions each element can provide, the more efficient your system becomes. For example, find a comfortable spot and carefully observe a tree. This tree, growing and blowing in the breeze, makes shade, turns carbon dioxide into leaves and branches, releases water vapor and oxygen, traps dust and soot, and may grow fruit, cones, or nuts. Planted strategically, a tree can block hot summer sun; provide mulch for the plants around it; cool surrounding outdoor air as it transpires water; provide sticks for trellis, fence posts, and bonfires; and encourage birds to fertilize the ground below. The art of thoughtful placement is key to integrative design.

As you arrange, keep relationships in mind. A water-harvesting basin could go under a permeable patio (with low plants growing between cracks in flagstone), next to a path so that the berm doubles as a raised path, below a chicken coop to trap nitrate-rich runoff for a fruit tree, or even on top of the roof. A rock-walled terrace can catch water, provide garden seating, support plants in soil pockets between stones, and soak up the sun's heat. Make the terrace bench height, and you could plant a frost-sensitive fruit tree below the stone terrace and next to the bench. The tree will tap the lens of water beneath the terrace's basin, stabilize the slope below the stone wall, and shade the bench. Radiant heat from the rocks will protect the tree from frost. Plant species such as yarrow and rosemary between the rocks to attract pollinators to the fruit tree.

Transform your landscape gradually. Start small, with one water-harvesting basin. Observe how it fills and drains and which plants thrive. Add more basins, terraces, ponds, or catchment areas. Tweak your design based on your observations. Repeat.

BUILDING BLOCKS AND THEIR FUNCTIONS

The building blocks of a sustainable landscape include trees and other plants, basins, soil, mulch, pavement and roofs, and cisterns. A successful integrative design will combine the building blocks with existing structures to provide all of these functions in a low-maintenance, beautiful landscape.

Trees provide shade, break the force of wind, and produce abundant mulch material in the form of leaves. Tree roots can tap water deep underground and stabilize steep slopes and the berms of rain garden basins. Like all plants, trees release water vapor as a byproduct of photosynthesis. This process of transpiration cools the surrounding air like a swamp cooler—but uses much less water. When planting a tree, think about how it will affect sun and wind reaching a building, garden, or patio.

Basins can range in size from shovel-sized pits that catch water and organic matter running down a hillside to curbside infiltration planters 6 feet (1.8 m) deep and the length of a city block. Long, narrow basins spread water out across the landscape. Berms can become raised pathways that shed water into sunken garden beds.

Soil filters pollutants and stores water and nutrients that plants use. Soil teems with life, housing invertebrates, fungi, and bacteria that decompose organic matter, fix carbon and nitrogen, and aerate the soil.

Organic mulch, such as leaves, dead grass, and sticks, shades the soil and reduces evaporation, decomposes into rich humus, and breaks the force of raindrops, which compact the soil. Covering bare soil with mulch is the simplest way to aid the infiltration of water and reduce the need for supplemental water and fertilizer. Stone mulch collects evening dew and provides all the other benefits of organic mulch except, of course, organic matter.

Pavement and roofs collect rain, and you can think of them as water sources. If you need water in a far corner of your landscape where water won't flow by gravity from the existing roof, consider building a patio or storage shed there and harvesting its runoff. Grade paved surfaces so water flows toward an area that needs irrigation. Collect runoff on top of the roof by growing a living roof there.

Cisterns are tanks that store rainwater. Water soaks up solar energy and stores a tremendous amount of heat. The thermal mass of a cistern can protect tender plants from frost and create a warm spot to sit on cool days. Tanks can be integrated into walls to provide a privacy screen or buried under patios or in basements. Tanks should be located where their overflow pipe can drain into an infiltration basin.

A sustainable garden is not just edible or attractive to wildlife, it is integrated into the home and provides outdoor living and cooking areas, seasonal shade and climate control (akin to passive solar design), shelter from wind, and privacy. As you consider expanding your garden's water-harvesting capacity with the strategies we discuss next, think carefully about where and how to integrate these elements into the existing landscape.

ADDITIONAL STRATEGIES FOR REDUCING RUNOFF

Here we describe several strategies that will allow you to catch more runoff than using a rain garden alone: reducing hardscape, harvesting rainwater in cisterns, and building living roofs, living walls, and curbside plant-

ers. Using these water-harvesting techniques is especially important in urban gardens or small landscapes, which may have limited space for rain gardens, but they work in large spaces as well.

REDUCING HARDSCAPE

In urban areas, your yard may be as large as your building or as small as the concrete patio where you keep your garbage cans. In either case, infiltration area is at a premium, so make every spot as absorbent as possible. The first step is to scrutinize your hardscape. Do you have more driveway than you use? If the patio is a concrete slab, consider making it smaller or removing it and rebuilding it with broken concrete pavers. Water can infiltrate in the cracks between pavers.

You have several options for making walkways permeable, depending on how you use the path. Main walkways should be 24 to 30 inches (60 to 75 cm) wide to allow easy walking, rolling, and wheelbarrow pushing. For wheelchair accessibility, use decomposed granite, pavers, or flagstone laid close together in sand or interlocking prefabricated permeable paving stones. In less-traveled areas or where accessibility is not required, use woodchips, brick, round river rock, or stepping stones. Woodchips decompose, adding organic matter to the soil, and can be obtained for free from tree trimmers. (Paths will need to be topped up with more woodchips every year or two.) Salvaged brick is available free or cheap from construction sites or salvage yards, and river rock may be available on site.

What about the driveway? If yours is a typical expanse of concrete, you have three options. Remove a strip of pavement down the center of the concrete driveway, leaving two tracks to drive on. Then turn the center strip into a rain garden or just plant groundcover. This option is

A RAINWATER-HARVESTING
CISTERN SYSTEM

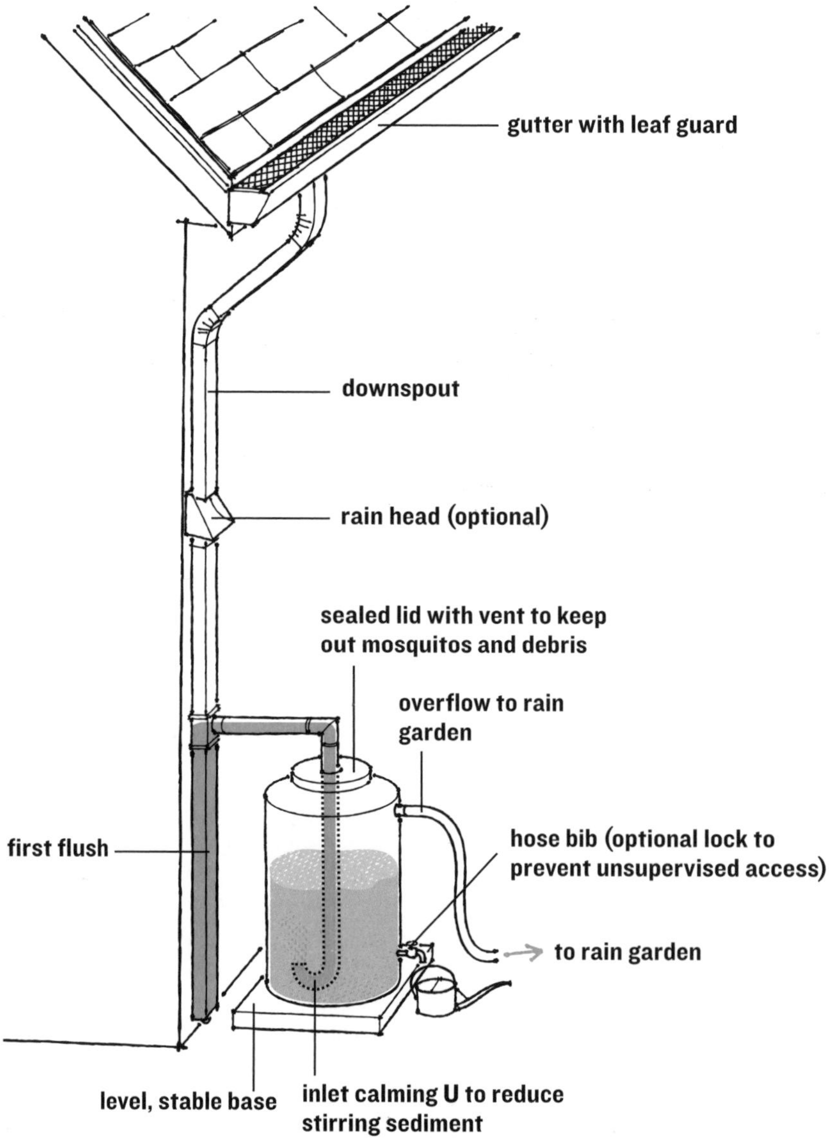

gutter with leaf guard

downspout

rain head (optional)

sealed lid with vent to keep
out mosquitos and debris

overflow to rain
garden

first flush

hose bib (optional lock to
prevent unsupervised access)

to rain garden

level, stable base

inlet calming **U** to reduce
stirring sediment

inexpensive—just the cost of renting a concrete saw if you do the work yourself. Use the concrete chunks for stepping stones, retaining walls, or patio pavers. Second, you could replace your driveway with permeable pavers. Check your local salvage yard first, because new pavers are not cheap. Finally, you could install a grass pavement system. This prefabricated plastic matrix supports the weight of your vehicle and keeps it from compacting the soil underneath. Install a well-draining soil mix beneath and in the matrix, then plant with drought-tolerant grass.

HARVESTING RAINWATER IN CISTERNS

When most people think of rainwater harvesting, they picture a rain barrel or tank hooked up to a roof. A cistern is a tank of any size, from the iconic 55-gallon (200-L) rain barrel to a 10,000-gallon (38,000-L) steel tank. Unlike rain gardens, which passively collect and use the rain, harvesting rainwater in cisterns involves plumbing and sometimes pumps. Because they store rainwater for irrigation (or even indoor use), rain barrels reduce overall household water consumption, thereby lowering your water bill and leaving more water in the river or aquifer that it comes from. Even a large tank will capture only a small fraction of the rain falling on a home roof. When the tank is full, rain will continue to flow into it from the gutters, then out the overflow pipe to the storm drain or, better yet, to a rain garden. Integrating both rain barrels and rain gardens into your home water system maximizes the benefit to your water source and local waterways. (Indoor rainwater use can significantly reduce dependence on municipal or well water but requires pumps and treatment, and this topic is beyond the scope of this book.)

Anatomy of a cistern system

A typical cistern system has five components: the catchment surface, typically the roof; the conveyance, the gutter and downspout or rain chain; the first flush diverter, a device that

CATCHING EVERY DROP OF RAIN by Christina Bertea

PLACE: Oakland, California, U.S.A

HOMEOWNER: Christina Bertea

DESIGNERS: Cleo Woelfle-Erskine and Christina Bertea

INSTALLED: January 2011

CATCHMENT AREA: 250 square feet (22.3 square meters)

RAIN GARDEN SIZE: 30 square feet (2.8 square meters)

ANNUAL RAINFALL: 22 inches (55 cm)

BY 2009, I had used my plumbing skills to retrofit all of my indoor fixtures with water-conserving models: a super low-flow showerhead that uses 1.25 gallons (5 L) per minute and a 0.8 gallon (3 L) per flush toilet. I was down to using a total of 11 gallons (42 L) of water per day. But what if the tap stopped flowing altogether, after an earthquake or extended drought? How would I supply my water? I needed to learn about rainwater harvesting, if only to have quality emergency storage on hand. However, I also wanted to store rainwater to help get my garden through the long dry summers of our Mediterranean climate and to keep winter stormwater runoff out of the San Francisco Bay.

My adventures in the rainwater world started with a one-day workshop. A salesman there gave me a little tank that he didn't want to ship back to New Zealand. Though it wasn't pretty, I welcomed the installation practice and invited friends over to watch me connect it. I built a dry system—the gutter discharges directly into the tank by gravity, leaving the downspout dry after the rain stops. Because the gutters sloped toward the front of the house, the tank had to go under the front downspout. Before the unsightly tank could upset the neighbors, I found an attractive 300-gallon (1500-L), 4-foot (1.5-m) diameter tank the color of sand. The overflow pipe now runs to a rain garden downhill from the foundation, which I got around to building a year later.

Instead of buying a kit, I fashioned a first flush setup from plumbing supplies. I attached a T fitting to the bottom of the standpipe that catches the dirtiest roof rinse water. A cleanout plug at the bottom of the T lets me empty out leaves and bits of fiberglass grit coming off the roof. On the branch I put a rubber cap with a hose bib (used for testing plumbing drain systems). When I turn the faucet on just barely, dirty water drips

153

CONTINUED

keeps dirt and bird droppings out of the storage tank; the storage cistern, a tank that includes an overflow pipe plumbed to the rain garden; and the distribution system, the hose or drip irrigation system.

As a catchment surface, the roof is the obvious choice because it is elevated, so water can flow by gravity into the storage tank and then into the garden. Driveways and lawns can also be used to collect rainwater to be directed to storage, but this option requires additional filtration and cost, as tanks must be buried and water pumped to the garden. You can harvest rainwater off most types of roof, including asphalt shingles, but beware of lead, copper, and fungicides in roofing or flashing materials. (Ask a roofer to identify your roof materials if you're not sure what your roof is made of.) These substances can contaminate your soil and get into food crops, where they pose a health hazard.

The conveyance system—the gutters and downspouts in the roof example—moves the water from the catchment surface to storage. Standard galvanized, enamel, or PVC gutters and downspouts work fine, but avoid using lead and copper, which leach heavy metals. Install gutter guards and downspout filters, referred to as rain heads, to keep leaves and large debris out of the cistern.

The first flush system ensures that water flowing into the cistern is clean. Between rains, catchment surfaces usually collect dust, airborne particulate pollutants, and bird droppings. During the first storm of the season, runoff carries these impurities off the roof. To keep potentially harmful substances out of the cistern, a first flush device is usually integrated into the conveyance line just before the storage tank. As rainwater flows toward the tank, the first flush device catches the first 25 gallons or so. After the first flush device is filled, rainwater flows into the cistern. A small tube at the bottom of the first flush lets the dirty water trickle into the ground.

The storage cistern must keep the rainwater clean and be able to withstand local conditions, including freezing and earthquakes. Cisterns should be opaque (to discourage algae growth), light-colored or shaded (to keep water cool and discourage bacterial growth), and made from food-grade materials. Cisterns also need to exclude mosquitoes.

Cisterns range in size from 55-gallon (200-L) barrels to 10,000-gallon (38,000-L) or larger tanks. Plastic, metal, fiberglass, wood, and cement tanks are the most common, but people also store water in plastic-lined bamboo frameworks, aboveground swimming pools, and underground cisterns made from plastered tires. New designs demonstrate creativity in fitting water storage into small spaces. Daisy chains of recycled 55-gallon (200-L) barrels can fit under a deck or in unused space between buildings. Pillow or bladder tanks fit in shallow crawlspaces under buildings or decks and inflate with water when it rains. Tall, narrow tanks made from corrugated steel culverts can be secured against a fence or in a narrow passageway. Whatever their design, storage tanks need to have a drain valve to send water to the distribution system and an overflow to divert runoff to a rain garden or storm drain for times when runoff exceeds storage capacity. In earthquake-prone areas, secure the tank to a wall or anchor it to a pad, and be aware that cement, stone, or brick cisterns may crack during an earthquake.

The distribution system carries water from the storage tank to where it will be used. Gravity flow from the tank works for hand watering close to the cistern, soaker hoses, and low-pressure drip irrigation systems. If you want to connect the cistern to an existing irrigation system, water plants far from the cistern, or send water uphill, you'll need to install a pump.

Cistern location

If possible, elevate the cistern or locate it up-slope from the plants to be irrigated, so that water can flow by gravity to the garden. This doesn't necessarily have to be next to the catchment roof—it can be anywhere in the landscape, as long as the inflow location is below the elevation of the gutter. A wet rainwater harvesting system takes advantage of the fact that water finds its own level, such that the buried pipe that runs between the downspout and cistern is always full of water. (In a dry rainwater harvesting system, the downspout leads straight from the gutter to the rain barrel, so the pipe empties between storms.) Locate the cistern on level, compacted, well-draining ground or on an elevated pad made from earth, stone, or stacked broken concrete.

Maintaining your cistern or rain barrel

Gravity-fed rainwater harvesting systems require little maintenance. Inspect all parts of the system annually, ideally before the rainy season's first large storm. Clean the gutters and downspouts, scrub out the cistern or barrel, check all fittings and connections for leaks, and inspect any other additional components of the system. Pumped systems should be inspected more frequently, as dirt and debris in the tank can increase wear and tear on the pump.

LIVING ROOFS

Living roofs replace conventional roofing with a living, breathing, vegetated roof system. The most basic living roof consists of a layer of vegetation over a lightweight soil mixture on top of a waterproof membrane. Any structure with a roof can support some vegetation.

Extensive roofs have 6 inches (15 cm) or less of soil, are lightweight (weighing 16–35 pounds per square foot [80–170 kg per square meter] when saturated with water), and utilize temporary or no irrigation. These low-maintenance systems are suited to smaller structures, such as a house, barn, shed, or chicken coop, and can be installed by the intrepid do-it-yourselfer. Intensive roofs feature up to 2 feet (0.6 m) of soil, a diverse plant palette, a permanent irrigation system, and usually require professional consul-

◀ **CASE STUDY CONTINUED**

out through a hose to nearby ornamental plants, emptying the standpipe within a few hours so it's ready for the next rain.

Then I decided to play with the tiny porch roof downspout, which obscured the lines of the decorative porch column. To get the water away from the house, I took an old, odd-angled downspout piece, aimed it out beyond the front of the house, and connected a length of rusty chain to the downspout using a spring. The chain leads to a vintage wooden rain barrel. I drilled a hole in the top, covered the hole with a fine screen to dissuade mosquitoes, and used a repurposed glass light fixture to guide the rain from the chain into the barrel. A hose bib at the bottom of the barrel lets the rain flow out a hose to the low point in that section of the front garden.

My ultimate rain-harvesting project will be to add a larger tank at the back of my yard to be filled using a wet system. The gutter downspout on my house will connect to a watertight pipe that will run underground, re-emerging next to the tank and rising high enough for the rainwater to enter the tank near the top. Because water seeks its own level, no pump will be needed. This wet system will allow me to position the tank far from the roof where the rain is being harvested.

In Mediterranean climates, it seems that the best way to minimize municipal water consumption is to utilize rainwater throughout the rainy winter (when the tanks can empty and refill repeatedly) for toilet flushing and, with UV sterilization, for clothes washing. With adequate filtration and testing, I could even consider using my rainwater as drinking water. In case of an emergency, I will certainly be out at the cistern with my little portable filter, feeling oh so prepared.

This wooden rain barrel receives runoff from a small porch roof via a rain chain, which is connected to the gutter with a spring.

155

tation, so we won't discuss them here. Brown roofs are an intriguing variation that makes sense if you have brick or concrete rubble on site. The crushed rubble becomes the growing medium for hardy plants, which provides habitat to ground-nesting bids. The roof initially lacks aesthetic charm, but soon attracts a wide array of insects, microbes, and lichens.

THE BENEFITS OF A LIVING ROOF

Although upfront costs are higher than traditional roofs, living roofs come out ahead when considering the longevity of the roofing materials. A living roof can also improve your quality of life and the local environment. Living roofs reduce runoff, create wildlife habitat, and can reduce nutrient pollution in local waterways by 30 percent by trapping water and dust that falls on rooftops. They reduce energy costs by providing insulation and summer cooling through transpiration. Living roofs are good for the neighborhood and the planet, because they help cool urban heat islands, absorb particulate matter and carbon dioxide, and reduce fossil fuel emissions from heating, cooling, and re-roofing. Plus, they look cool, spark conversation, and can last for fifty years.

Structural requirements

If your roof slopes less than 40 percent, you

LIVING ROOF LAYERS

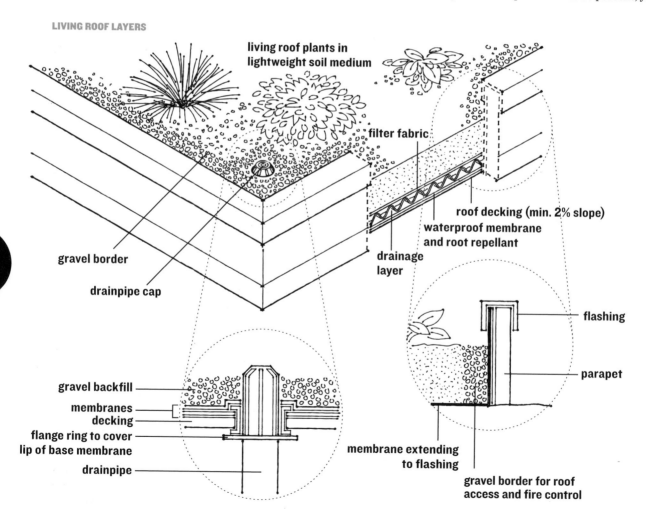

living roof plants in lightweight soil medium

filter fabric

roof decking (min. 2% slope)

waterproof membrane and root repellant

drainage layer

gravel border

drainpipe cap

flashing

parapet

gravel backfill

membranes

decking

flange ring to cover lip of base membrane

drainpipe

membrane extending to flashing

gravel border for roof access and fire control

THE ROYAL BANK OF CANADA
RAIN GARDEN by Nigel Dunnett

PLACE: London Wetland Centre, West London, U.K.

DESIGNERS: Nigel Dunnett

INSTALLED: Summer 2010

RAIN GARDEN SIZE: 3600 square feet
(335 square meters)

ANNUAL RAINFALL: 23 inches (58 cm)

THE RAIN GARDEN at the London Wetland Centre was opened in September 2010, as part of the celebrations for the tenth anniversary of this important nature site in West London, owned by the Wildfowl and Wetlands Trust. The London Wetland Centre was created on the site of former reservoirs and water treatment works. The center is visited by tens of thousands of school children every year, together with thousands of families and individuals who come to learn about wetland wildlife and to enjoy the beautiful landscape and extensive visitor facilities.

The aims for the garden were very clear: to provide an interactive and very attractive garden that would demonstrate the importance of water conservation and inspire people to try out water-saving and rain garden techniques themselves at home. I wanted specifically to use the garden to introduce and promote the concept of rain gardens in the United Kingdom, as until then there had been very little application of the idea. Unlike other existing garden areas at the London Wetland Centre, the intention was that the rain garden be fully accessible to visitors so that people could investigate and see how things worked. Therefore, the construction, materials, and planting had to be very robust. Above all, the idea was that people would enjoy the garden and say to themselves, "That's fantastic, I can do that at home!"

The garden space measures 60 feet by 60 feet (18 m by 18 m) and has an existing small stream running through the middle. The land rises gently on each side of the stream by around 3 feet (90 cm) to create a very slight valley landform. The layout of the garden is loosely based on a series of concentric circles, suggesting radiating ripples created by rainfall on a pool. A widely curving boardwalk brings people through the garden, and there are many small exploratory paths. Stepping stones enable brave people to cross the stream: It was very important to give as much interaction with water as possible, to counter the pervasive view that open water in the landscape is a danger.

At the heart of the garden is a series of circular raised beds that lead off from a small garden building that acts as a shelter and information point. The beds, which step down a gentle slope, soak up water that is passed along a series of wooden rills. Each mini-garden absorbs some of the water into the soil, but much of it is drawn up and transpired by the planting, sending it straight back into the atmosphere. Any excess runoff overflows and moves onto the next bed. The planting in each bed is slightly different, responding to the decreasing moisture gradient that is produced. One of the beds is filled with water rather than soil and planted with true aquatics (which have a water-cleansing function), while the rest employ moisture-tolerant species, carefully chosen to survive the wide-ranging conditions they will experience, which can include extended periods of drought.

Ultimately little if any water is left to pass into the stream. Consequently a treadle pump has been installed to lift water up from the stream, allowing it

157

CONTINUED

can convert it into an extensive living roof. Before transforming the roof of your house, we encourage you to experiment on a shed or chicken coop or seek out a local workshop or professional consultation. Make sure the foundation and structure can support the saturated weight of the soil, and attach the membrane and drainage scupper carefully.

Before you consider building a living roof, get up on the roof and make sure it feels strong and sturdy. If the roof supports a snow load without collapsing, it should support an extensive living roof. If in doubt, consult an expert. Water weighs 62 pounds per cubic foot (1000 kg per cubic meter). A thorough analysis of the roof structure may also reveal areas where point loading can be increased—perhaps over a column or along a bearing wall—to hold deeper soil and larger plants.

Living roof layers

The waterproof membrane keeps water out of the structure below the roof, and many types of membranes can be used. Pond liners are great, and other materials can be used as well; consult a local professional for locally available membranes. They come in many thicknesses, and the thicker the better. Leaks can cause severe damage, so pay close attention to membrane installation. Consult a local roofer for assistance.

A drainage system performs two functions. First, it allows the soil—a specialized, light-

Suzanne Forsling affixed gutters to the sunny side of her home and made a living wall of vegetables.

158

159

to fall back down the lower rills. Working the pump engages visitors in a physical activity and demonstrates what would otherwise be rather a subtle effect, as under normal conditions no rainwater should appear at the lower end. The intention with the pump is to show visitors exactly how rain gardens work: a case of making the water cycle visible.

The focus of the garden is an ingenious building that collects rainwater and supplies it to the garden. This rain shelter is a converted shipping container with a diversely planted green roof. The container had a previous life, moving cargo around the world, and is a prime example of recycling and reuse of materials, a concept underpinning the use of materials throughout the whole garden. While building the garden it became a matter of principle to source reclaimed materials wherever possible; this not only helped stretch the budget, but also produced a highly individual garden and one from which people leave inspired by ideas that they can easily employ at home. This can perhaps best be seen in the constructions of the walls around the rain garden areas, but also in the dramatic and

Each circular rain garden will fill with water after a heavy rain and then overspill to the next in the chain. The third garden contains open water and ornamental reeds to help cleanse the water as it passes through.

CONTINUED

This addition to a home by Bricault Design includes three sides of living walls.

imposing towers that rise up from the wildflower meadows that fill the garden. These are multistory habitat and feeding sculptures that have the potential to attract a wide range of life. The planting throughout the garden is naturalistic and meadow-like in character.

The rain garden has been an outstanding success. The combination of wildlife-friendly features, colorful planting, the curving arc of the linked rain gardens, and the striking nature of the green-roofed building produce an impression that is quite different from the normal garden experience. "From day one, the garden has been full of children and adults enjoying the interaction with water and nature, and they all come away knowing what rain gardens are all about. That's exactly what we wanted," says Simon Rose, the Developments Manager for the Wildfowl and Wetlands Trust.

All excess rainwater runoff from the green roof runs down a rain chain into a cistern, which overflows via a channel into the first in a sequence of rain gardens. The end of the building is clad with panels that provide habitat for garden wildlife.

CASE STUDY

STREET ORCHARDS FOR COMMUNITY SECURITY

by Brad Lancaster

PLACE: Tucson, Arizona, U.S.A.

DESIGNER: Brad Lancaster

INSTALLED: Ongoing

ANNUAL RAINFALL: 12 inches (30 cm)

SWISS ecovillage designer Max Lindigger's story of a walk he took with his grandfather radically changed my view of public streets. His grandfather pointed to condominiums sprawling above his alpine village, "That's where we grew and gathered food during the war. The forests were common land. What commons remain?"

I looked at my Sonoran Desert city of Tucson, Arizona, and asked myself, "Where are my community's forests, our commons? Where would we get food in times of need?"

More than 450 native food plants grow wild in the intact areas of the Sonoran Desert, many of them reducing the occurrence or effects of diabetes. The velvet mesquite tree (*Prosopis velutina*) is one such keystone species, producing naturally sweet protein and carbohydrate-rich seedpods in both wet years and dry. No wonder it was a staple of Yaqui and O'odham diets. Within my city, almost all mesquite forests have been

replaced with hot and inhospitable pavement or thirsty exotic plants. The pavement that drains much of our scant annual rainfall is also how most of our food arrives. If oil supplies fueling semi-trailers disappeared, we'd be without food. If the power that fuels our well pumps went out, we'd be out of water. We are creating the conditions for catastrophe. But that can change. The majority of public land—our commons—in the urban setting is our public streets and adjoining right-of-ways. The resources (soil, native plants, rainwater runoff, and people) to grow a food forest on these barren streetscapes are in our own neighborhoods.

Once established, native food plants can survive without irrigation. When irrigated with harvested rain, they can thrive. Because of the sun-baked caliche soils, almost all of Tucson's stormwater runoff flows straight from roofs, driveways, patios, and parking lots to public streets that flood to resemble rivers; the runoff then exits via storm drains—

a huge waste that creates a whole set of water management problems as well.

Desert Harvesters, a collaborative effort in my hometown, is helping turn that waste into a resource by diverting stormwater runoff to grow indigenous, food-bearing shade trees in curb strips and in-street traffic-calming islands, then showing people how to harvest and process the bounty. We organize neighborhood food tree planting parties and show neighbors how to plant native trees in water-harvesting earthworks along streets and in their yards. (Local partners provide the velvet mesquite trees at a discount.) The planting parties bring neighbors together to beautify their blocks, build community, and provide public examples of water harvesting strategies.

Within six years, the trees provide shade and colorful seasonal blossom. We find that as native habitat grows back within the urban core, native cardinals, flycatchers, cactus wrens, hummingbirds, curve-billed thrashers,

CONTINUED

A mix of native and ornamental plants thrive in this garden, which is watered with greywater from a kitchen sink.

Chi Lancaster eats mesquite harvested from a community street orchard planted ten years earlier by the Girl Scouts and watered with runoff entering through curb cut on lower right.

white-winged doves, Gamble's quail, and Gila woodpeckers replace pigeons. Within eight to ten years of planting, the tree-shaded sections of the neighborhood are noticeably cooler (up to 10°F [6°C], according to studies) than unplanted areas. Plant a tree and you plant a living air conditioner.

We live in a society that is often short on time and in search of convenience. Traditional hand-grinding of mesquite pods and processing other wild foods takes many hours. We sought to speed up the process and make it fun, so we bought a farm-scale hammermill and mounted it to a trailer. We take the mill to public places, and people bring their harvested mesquite pods to grind. The hammermill can

CONTINUED

weight growing medium—to drain so that plant roots don't rot. Second, it sends water off the roof quickly during heavy rains so water does not pool and damage the roof structure or membrane. Commercial living roof systems use expensive plastic drainage layers, but you can improvise one by drilling holes in recycled fiberglass roofing. Lay geotextile fabric over the fiberglass to keep the growing medium from falling down into it and clogging the drainage holes. An even simpler solution is to create gravel channels that are approximately 8 inches (20 cm) wide and the depth of the growing medium in between the planting beds.

As an overflow drain, you'll need a scupper, a metal or plastic spout that sends runoff to a gutter or rain chain. The scupper goes through the membrane. If a living roof leaks, it will most likely leak at the scupper, so construct this detail very carefully, or hire a roofer for this small job.

The growth medium must support, water, and fertilize plants, all in less than 6 inches (15 cm) of material. This trick is accomplished by provide a mixture of topsoil, compost, and inorganic material. Topsoil (sometimes supplemented with peat or coconut fiber) helps retain water, compost provides nutrients, and inorganic lava rock or pumice provides pore space for aeration and drainage. Start with a mixture of 1 part clean topsoil, 1 part compost, and 1 part perlite, lava rock, or crushed brick. Add a layer of coarse mulch on top to retain moisture and to keep growth medium from drying out and blowing away.

Living roofs are hot, windy environments, and most plants wither and die without irrigation. But succulents such as *Sedum*, *Delosperma*, and *Sempervivum* species thrive in these harsh conditions, and they stay green year round and resist fire. Grasses and forbs with fibrous roots (rather than taproots) can also do well. Investigate local sand dunes or rock outcrops for plant ideas—the ones you find will likely do well on your roof. Avoid plants with tap roots, as they can puncture some types of membrane. Water plants until they become established, and remove weeds annually. Don't be alarmed if some of your plants die, and be prepared for natural succession.

Additional design considerations

If your roof slopes more than 6 percent, watch for signs of erosion. Replace mulch if you notice soil washing or blowing away. On roofs with a slope greater than 30 percent, install jute netting over the mulch, and plant through the holes.

To encourage maximum biodiversity on your living roof, design a mosaic of vegetation of different heights. Add small mounds and berms to provide sunny and shady microhabitats for invertebrates and microorganisms. You can install solar panels just above the surface of your living roof, and the shade they cast will create cool microclimates.

The soil layer also acts as a fire barrier, but it is still a good practice to choose plants that are slow to ignite. Succulents, which store water in their stems, are good choices, whereas ornamental grasses become kindling in dry conditions. A 12- to 24-inch (30- to 60-cm) perimeter of gravel or pumice will act as a fire break and also provide maintenance access to the roof.

LIVING WALLS

Don't have access to a roof available for planting? Don't worry. Vertical walls are another great surface to cover with plants. Living walls, also known as biowalls, vertical gardens, and sky farms, can grow inside or outdoors. Just like living roofs, living walls offer a cornucopia of benefits, including better insulation, increased wildlife habitat, food production opportunities, cooling, and improved air quality. Living walls range from do-it-yourself to highly engineered and artistically designed projects, and they complement rain gardens and cisterns. For example, water pumped from a cistern through a drip irrigation system can irrigate living wall plants growing in light-

▶ **CASE STUDY
CONTINUED**

Brad Lancaster harvests all the rain that falls on his roof and landscape. The velvet mesquite orchard along the street harvests rain that runs off the street.

grind 5 gallons of whole mesquite pods into I gallon of finely textured, naturally sweet flour in just 5 minutes. The flour sells for $14 per pound and up. So harvesters can easily make $25 per hour harvesting, processing, and selling the flour in their own neighborhoods—especially at our annual mesquite pancake breakfast and milling. Desert Harvesters published *Eat Mesquite: A Cookbook* to share the pancake recipe and delicacies such as mesquite baklava, ice cream, syrup, mole, pizza, crackers, bread, cookies, and much more.

RESOURCE

For more on Desert Harvesters and a link to their cookbook, see DesertHarvesters.org. Brad Lancaster's book, *Rainwater Harvesting for Drylands and Beyond*, Volume 2, and website, www.HarvestingRainwater.com, show many other ways to "First plant the rain, then plant food trees."

weight hydroponic mats. Even more simple, water splashing off a flat stone or sculpture on its way to a rain garden can passively water drought-tolerant succulents. Simpler still are stone walls or terraces that feature plants growing out of the crevices. Here we summarize types of living walls and present ideas for the ambitious vertical gardener.

The classic living wall features climbing ivy or trellised vines and is also known as a façade living wall. Plants grow from soil at ground level, so species are limited to vines, which can take years to mature and cover the wall. One advantage to this type of wall is that a downspout can be easily directed to the ground soil for these wall climbers, with overflow going directly to a rain garden. This eliminates the necessity of irrigation systems. Problematically, aerial roots tend to attach to the wall and can cause structural damage.

As designers explore the relationship between gardening, water conservation, and energy conservation, new modular approaches to vertical gardening are emerging. Soil systems feature an array of soil-filled cells, and these are usually some kind of tray or bag affixed to a framework attached to a waterproof protected wall. Bringing the soil up the wall dramatically reduces the time needed to create a seamless wall of leaves and flowers. It also broadens the plant palette from vines to succulents, ferns, and other plants with slow-growing roots. One advantage of a modular soil system is that dying plants are easy to replant. That's a good thing because the small compartments in some modular systems limit growth and nutrient availability. When plants become root bound and die, both plant and soil need to be replaced.

Hydroponic soil-less mat systems feature plants growing through water-holding mats (usually made of felt). Plant roots form a strong matrix along the back of the mat, which is backed by another piece of felt and a waterproof membrane. These panels are secured to framework in front of the wall. Without soil, this type of living

wall is very lightweight but requires more water to keep the plants healthy. Mist or trickle irrigation delivers water through the mat system. Water can be captured in a trough at the base of the wall and recirculated.

Many living walls will require regular irrigation. Drip line is an efficient way to get water to the plants without reducing aesthetic appeal. Whenever water is applied near a building wall, a waterproof backing or shield is installed between the living components and the wall to ensure that no water can permeate the wall. Drips and runoff must flow away from the wall. Airflow is also necessary for healthy plant growth. Make sure the growing medium is not too dense and that space exists between the wall and plants.

Living wall maintenance is fairly simple. Regular tasks include occasional removal of dead leaves, thinning plants, and light weeding. Check soil or growing medium and plants annually, and replace soil or replant as necessary.

Living walls are a playful and creative way to add interest to a space, and they don't have to be large or high tech. Start small, with a living wall planter or stacked planter boxes. Look around for interesting found objects, plant them, and hang them on the wall. You'll be on your way to an eclectic and noteworthy vertical garden.

CURBSIDE PLANTERS

The last place to infiltrate runoff before it hits the street is the curb strip—the long, skinny area between the sidewalk and the curb. In dense urban areas, this may be the only patch of unpaved ground that can receive the rain from your roof. (Some cities pave all the way to the street, but may allow you to remove some pavement for tree planting.) If your downspouts let out at the sidewalk, check to make sure that water flows away from the building and toward the street. Then dig a basin in the curb strip. Because the curb strip is usually less than 10 percent of the roof area, a 2-foot (60-cm) deep rain garden won't absorb all runoff. You can expand its capacity by building a dry well under the rain garden.

If you don't need the curb strip to catch water from your roof or driveway, you can use it to infiltrate water that runs down the street. (Cutting into the curb typically requires a permit from the public works department.) First, figure out which way water flows down your street. Even flat-looking streets have some slope. Remove two short sections of the curb, each about 16 inches (40 cm) long, at the uphill end of the curb strip. Then dig a basin in the curb strip. A small mound of dirt below the curb cut intercepts water flowing down the gutter and diverts it into the basin. The dirt mound will need to be replenished periodically, so if your area permits, you could ask the city to install a speed bump diversion swale. When the basin fills, it overflows through the curb cut back into the street, leaving sediment and heavy metals behind. Because street run-off contains copper, zinc, motor oil, and other contaminants, don't use the curb strip planter to grow edible plants. Trees with aboveground edible fruit or nuts are okay, though, because most metals get bound up in the wood.

WATER-WISE IRRIGATION

Although your rain garden should need little irrigation, vegetable beds, ornamentals, and fruit trees do need extra water. Here we introduce a suite of water-conserving gardening strategies and show how to integrate them into your home landscape. Each of these concepts merits a book of its own, and we've listed several such titles in the Resources section.

USING GREYWATER IN THE GARDEN

If you live in the United States, you use about 50 gallons (190 L) of water per day to bathe and wash dishes and clothes. This greywater—so called because soap and grime tint it grey—

is great for watering plants. If everyone reused their dirty bathwater, our cities would remove a third less water from rivers and aquifers and reduce the amount of sewage we treat by 60 percent.

It's easy to reuse your greywater in the garden. First, dig a circular trench encircling the stem to a depth of 9 inches (22.5 cm) around fruit trees, shrubs, or large annuals like tomatoes. Fill the basin with woodchip mulch. These basins keep greywater from running into a pathway or neighboring property, and keep children and pets away. Mulch keeps grease and soap from clogging the soil, where dirt, soap, and any bacteria break down.

Now figure out how to get the water from your shower, sink, or washing machine to the garden. You can capture shower water in a 5-gallon (20-L) bucket placed in the shower and collect sink water by placing a bucket under the (disconnected) sink drain, then carry it outside and pour it into the mulch basins. Or ask a plumber or handy friend to divert your drainpipes outside. Simple greywater systems are legal in Australia and several states in the United States, and cost $75 to $200 if you do the work yourself.

Once you disconnect from the sewer, your landscape becomes your water treatment plant. Avoid toxic cleansers, and never use the drain as a dumping ground—the sewage treatment plant doesn't remove toxic chemicals, and you certainly don't want them in your garden. Sodium and boron are fine for us, but bad for plants and soil, so buy liquid detergents without these ingredients. Then load up the washer and watch your garden thrive.

The simplest greywater system: an outdoor shower

If plumbing seems daunting, make an outdoor shower. They require little plumbing, use the sun to heat water, and provide an irrigation boost in the summer when plants need it most. Best of all, they're legal even where conventional greywater systems are not. Because they need no drainpipe, outdoor showers are technically not regulated under the U.S. Uniform Plumbing Code.

To get the water hot, you can use a manufactured hot water panel, a spiral of black polyethylene irrigation tubing or copper pipe, or a black 20-gallon (75-L) food-grade barrel. Most outdoor showers use city water pressure to push water into the water heater. The simplest method is to just hook up a hose. If you have no city water pressure or just want to impress your friends with you physics acuity, make a thermosiphon, which uses the fact that hot water rises to draw water up into the tank.

The simplest way to drain your outdoor shower water is to dig a diversion swale under the shower, fill it with river rock or cobble so greywater doesn't surface, and build a wooden deck on top. A repurposed palette makes an adequate deck. You can also skip the deck and stand right on the river rock. Just make sure the bottom of the diversion swale has at least a 4 percent slope so that the water drains quickly. Direct the swale to a mulch-filled basin planted with your favorite plants or to an existing tree or garden.

If you have more than one area you would like your water to nourish, integrate multiple drains in the shower floor. Connect each drain to a pipe leading to a desired tree or garden. Place caps over all the drains but one. Each time you shower, rotate the drain caps so a new area is getting watered.

Another possibility is to grow the shower enclosure. Grade the ground under the outdoor shower so that it is highest in the center and slopes gently away to a circular trench. Plant bamboo, willow, reeds, ocotillo, or other tall, fast-growing plants in this trench, and the shower will water them. If you choose willow, you can weave the branches together, then scrape bark from the branches where they touch and lash them together with a rubber band. The branches will grow together, eventually forming a living wall.

WAYS TO SAVE WATER IN THE VEGETABLE GARDEN

Sheet mulch gardening is a way to build an instant garden on top of the soil (or lawn), rather than digging it up. After disassembling your vegetable garden in autumn, add compost and plant a cover crop. Winter moisture will sink into the soil and fill spaces between the soil particles. About a month before planting the garden in spring, mow the cover crop and cover the bed with a layer of cardboard topped by straw or leaf mulch. Cut small holes cut in the cardboard, and plant seedlings or large seeds in pockets of topsoil. Alternatively, you can till in the cover crop and plant vegetables or flowers.

Add mulch while you dig the garden beds. Mulch keeps soil moist and loose, suppresses weeds, and encourages helpful soil critters. Potatoes, garlic, and large seeds will grow through thick straw or leaf mulch. For small seeds, mulch lightly with straw or leaves until the seeds sprout, then add more straw between the plants.

Use low-tech drip irrigation. Unglazed terra cotta pots buried in perennial beds or container gardens will wick water to deep-rooted plants. Use a narrow-necked pot and plug the drain hole with clay. Choose pots with a 2- to 3-gallon (7.5- to 10-L) capacity. Fill the pots with water using a hose, then cover the top with a piece of stone or tile. Water will seep slowly through the pores in the clay and into the soil. Depending on the climate, you'll need to fill the pots once or twice a week.

Many crops can be dry farmed—grown with only the rain—even in dry climates. Nothing is sweeter than a dry-farmed peach or 'Early Girl' tomato. Plant tomatoes, squash, or melons 6 feet (1.8 m) apart and mulch well.

" LAST WORDS

REMEMBER THAT the most successful home water systems evolve gradually. We recommend approaching rain gardening with a patient and adaptive spirit. As a recent rainwater harvesting study from India pointed out, humans have been harvesting the rain for more than 8000 years, through a process of constant tinkering in response to changing needs and climates. From this perspective, if you start small, expand slowly, and constantly refine your designs along the way, you'll end up with a beautiful, functional, and largely self-sustaining landscape that benefits your family, your neighborhood, and your local waterways.

COLLABORATIONS ON COMMON WATERS

IN THIS EPILOGUE, WE SHIFT OUR FOCUS

from the home landscape to the watershed. We show some ways that rain gardeners have connected the individual act of rain gardening to salmon recovery, oyster cultivation, and urban food production. Through these brief stories, we hope to inspire you to join millions of people around the world who are creating new water cultures.

To develop new water cultures, we need to find new ways to build, grow food, make electricity, get from place to place, and manage wild lands. If you worry that one rain garden's effect on the many problems facing the world's waters is insignificant, take heart. Around the world, people are putting shovel to soil this very minute, replanting vanished forests in their landscapes or along a local stream. Many more people are coming together to farm sustainably, conserve energy, restore oil-soaked wetlands, and remove dams. You might join them tomorrow or you might have already done so long ago. Either way, your rain garden can serve as a touchstone and daily ecological collaboration.

By taking this book's message into your garden, your rain garden will become a living sponge that slows the flow of water across the landscape and filters contaminants before they reach the nearest beach. We hope that by designing, building, and living with a rain garden, you've gained new appreciation for your neighborhood's hydrology and the opportunities to reroute its stormwater flows. Rain gardens show how people can inhabit places in a way that makes ecological sense. This is a rain garden's greatest boon: the insight that living near a swamp or stream doesn't have to mean despoiling it. Rain gardens change the way we inhabit our spaces and can bring new perspective to our place in the water cycle.

Rain gardeners tell us that having a rain garden reminds them daily that every drop of rain eventually reaches a waterway, along with the pollutants it picks up along the way. This awareness can percolate outward, leading people to commit to change other destructive habits. A rain garden shows that if we deal with our impact where we live, there's no need to offset it somewhere else. Instead of feeling guilty about your lifestyle's harmful impacts, figure out how to expand your circle of beneficial influence.

Collaboration is the key to restoring damaged landscapes. Humans can collaborate with

PREVIOUS SPREAD:
The new Bowens Island restaurant sits just above the high tide line. In the foreground, oysters have colonized an experimental artificial reef.

water, plants, animals, and the land for mutual benefit. A rain garden is one such collaboration among the rain gardener and the plants, worms, fungi, and microbes that help infiltrate water and absorb nutrients. Larger collaborations—involving landowners, fishers, government agencies, and ecosystems—often increase economic opportunities, restore ecological balance, and create green space and social connections. Near New Orleans, shrimpers and environmentalists helped craft a plan to treat sewage in constructed wetlands because the wetlands shelter young shrimp and offer hurricane protection. On the Klamath River, tribes, farmers, and fishers in southern Oregon and northern California support removing four hydroelectric dams because the dam removals will increase failing salmon runs and create fishing and recreation jobs. One of the longest and most successful watershed restoration collaborations is taking place in the Nisqually River basin in northwestern Washington, where rain gardens are the latest tool in a 25-year effort to restore salmon.

EXPANDING YOUR POSITIVE IMPACT

▶ Using nonsynthetic fertilizers reduces dead zones in your local creek and the ocean.

▶ Planting an edible garden reduces irrigation elsewhere, as well as the need to transport food over long distances.

▶ Using wood from a sustainably managed forest for a building project keeps sediment out of salmon streams.

▶ Conserving energy and supporting solar and wind power close to home reduces dependence on hydroelectric dams and fossil fuels.

▶ Replacing water fixtures with low-flow models keeps more water in lakes and streams.

NISQUALLY: RENEWING AN ANCIENT COLLABORATION WITH SALMON

The Nisqually River flows from the slopes of Mount Rainier to the geoduck sands of Puget Sound. There, collaboration among local tribes, landowners, town planners, and government agencies is transforming the watershed and the ways that people think about their place in its ecological fabric.

The Nisqually is becoming known as a model for collaborative watershed manage-

Nisqually native plant technicians install tree protector tubes on some of the native trees and shrubs their crew has planted along Ohop Creek, an important salmon stream.

ment, but the river is most famous as the flash-point of the fish wars of the 1970s. Tribal fish-ers up and down the west coast of the United States staged fish-ins, civil disobedience pro-tests against racist fish and game regulations that made tribes' salmon fishing illegal. Back in the 1850s, Nisqually treaty negotiator Les-chi insisted on retaining the riverside prairies in the lower Nisqually watershed and the right to take fish "at all usual and accustomed grounds and stations . . . in common with all citizens of the Territory." Although the 1857 treaty and subsequent land grabs eroded the Nisqually land base, many Nisqually made a good living from fish and shellfish through the 1950s. The post–World War II population boom created a lethal brew of hydroelectric development, pesticide runoff, and out-of-control commercial fishing that sent salmon populations into freefall. State game wardens blamed tribal fishers and waged an increas-ingly violent campaign of arrests and sting

spawn in the river or are eaten by bears or seals). The Boldt decision also recognized that tribes had a legal right to manage natural resources on equal footing with state and federal wildlife agencies.

The approach to co-management that arose on the Nisqually River demonstrates the tribe's resourcefulness and creativity in protecting the fish, the river, the land, and the Nisqually people. Like other tribes of the Pacific Northwest, the Nisqually formed natural resources departments, hired biologists, and adopted fishing ordinances. Billy Frank Jr., a veteran of the fish-ins, became the Nisqually Tribe's fisheries manager and chair of the Northwest Indian Fisheries Commission, a project of nineteen western Washington tribes. The commission meets annually with the state to set fishing limits and decide on management activities, and it employs about fifty biologists, ecologists, computer modelers, and lawyers to craft and defend the tribes' management plans.

The tribe's goal is to restore a subsistence and commercial fishery to the Nisqually River, which entailed tackling hydroelectric dams, farming, and logging. The tribe developed an ambitious salmon recovery plan that affected land and water use from the headwaters on Mount Rainier to the delta on Puget Sound. Since the Boldt decision, the Nisqually Tribe has returned to court to litigate shellfish, levee, and culvert issues but has become equally adept at negotiating collaborative solutions that benefit everyone in the watershed. These negotiations take place at monthly meetings of the Nisqually River Council, which bring together local government, state and federal resource agencies, the tribe, the Tacoma power company, and landowners in the watershed.

Since its formation in 1987, the Nisqually River Council has protected riverside lands, restored tidal flow to the estuary, created dynamic vegetated floodplains to straightened sections of river, and negotiated changes in dam operation to create more natural river flow. Nisqually elder and natural resource manager

operations along rivers and lakes across the United States.

The Nisqually and Puyallup Rivers became the focal point of treaty-rights challenges to state fishing regulations. After a ten-year campaign in the courtroom, in the media, and along the rivers, the tribes prevailed in 1974. In the watershed case *United States v. Washington*, Judge Hugo Boldt affirmed tribes' rights to 50 percent of the harvestable salmon. (Nonharvestable salmon include those that

CASE STUDY

RAIN GARDEN CLUSTERS: SUSTAINABILITY AT THE NEIGHBORHOOD SCALE

PLACE: Eatonville, Washington, U.S.A.

HOMEOWNERS: Six families

DESIGNERS: David Hymel and Marilyn Jacobs, Rain Dog Designs LLC

INSTALLED: 2010

RAIN GARDEN SIZE: Each ranges from 120 to 150 square feet (11 to 14 square meters)

ANNUAL RAINFALL: 43 inches (108 cm)

ADDRESSING stormwater pollution in Eatonville is a critical part of salmon habitat restoration in both the Mashel River and Ohop Creek, which run on either side of the town. It's also one small step toward cleaning up Puget Sound, which is contaminated by industrial waste and urban runoff, and its orcas, which are among the most contaminated marine mammals in the world. The bulk of Eatonville's stormwater is directed away from the Mashel River and sent untreated into Ohop Creek. The Mashel River has low flows in the summer and early autumn, which make the river too warm for young fish and too low for adult salmon to get upstream. Rain gardens infiltrating stormwater into the groundwater will help increase base flows in the Mashel, which will keep the river cooler and make it easier for adult salmon to travel from the Nisqually to their spawning grounds in these tributaries.

In August 2009 Stewardship Partners piloted a novel approach in its low-impact development outreach program by engaging an entire neighborhood to install several rain gardens in a cluster in Puyallup, Washington. Building on this, planning began in January 2010 for a second cluster in Eatonville. Up to this point rain garden efforts had been focused on single projects, and these would be the first multiple installation retrofits for private homeowners in Washington. The aims were to reveal the hurdles to installing rain gardens in an already developed residential area and to see what efficiencies could be gained installing a larger number of rain gardens at one time. The hurdle in Eatonville proved to be the same as in Puyallup: finding a neighborhood and homeowners willing to host free rain gardens in their yards with the condition that they are on adjacent parcels. There was general interest, but not enough adjacent homeowners had responded to create a cluster demonstration. A breakthrough came in March, when we contacted a homeowner on Baumgartner Avenue North who was already enthusiastic about rain gardens through her experience with a project at an

This rain garden planting event brought together neighbors, friends, and community.

176

CONTINUED

Georgiana Kautz says that by working together and showing respect for each other, council members have realized they have a common goal: to leave clean water, healthy forest habitat, and salmon for future generations.

To bring the whole watershed along on the long road to salmon recovery, the council sponsors an eight-week course on the watershed every summer. The class is one way the Nisqually communicate the perspective they have gained through at least 12,000 years of management to newcomers to the watershed. The class is free, but participants must volunteer 40 hours on any project related to watershed health, from willow planting to making paintings of local landscapes that the council can use for education.

The watershed class played a key role in starting the council's rain garden program. Local landowner and contractor David Hymel took the class in 2006. While completing his required volunteer hours, he sent an email to the Nisqually River Council mentioning that he would like to bring together lending institutions, engineers, real estate agents, and project owners to promote low-impact development. The email reached Stewardship Partners, a Washington nonprofit that worked with the council to expand its programs. Stewardship Partners hired Hymel to help promote low-impact development in Eatonville, set in the middle of the Nisqually watershed at the confluence of two tributaries, the Mashel River and Ohop Creek. The council's geographical analysis showed that these streams could provide prime salmon habitat if restored, but rapid growth in Eatonville threatened water quality and the success of in-stream habitat structures.

Enter the rain gardens. Jeanette Dorner, tribal salmon recovery coordinator, believes that Stewardship Partners' rain garden campaign marked a turning point, because rain gardens made a clear connection between riparian plantings along Ohop Creek, water quality, and salmon recovery. Sally King lives in Eatonville and organized six neighbors to participate in the rain garden program. She says she began with selfish motives of dealing with drainage problems that left her lawn brown in patches. But by building and living with a rain garden, she has realized that it is also good for the river. King has since gone on to expand her positive impact by replacing part of her lawn with a permeable walkway and installing a rain barrel.

Eatonville mayor Ray Harper was so excited by rain garden possibilities that he made sure the new town square featured rain gardens and permeable pavement that lets the rain sink in. The town is now figuring out how to get rid of the storm sewer entirely, through a combination of rain gardens, permeable pavement, and cisterns. Such a project will not just reduce the negative impact the town has on the salmon, it will expand the positive impact of development by creating wildlife habitat and green space within the town limits.

Today things are looking up for the Nisqually River salmon. With tidal flow restored to the estuary, young fish are growing larger and stronger before they swim out into Puget Sound. When they return to spawn, salmon find three-quarters of the riparian area protected and more shady backwaters to spawn in every year. The Nisqually collaborations show how rain gardens can contribute to watershed restoration and how vibrant watersheds support healthy communities and sustainable economic development. Each rain garden cluster that Stewardship Partners installs makes the water in the river a little bit cleaner and cooler and builds community support for long-term initiatives like community forestry and sustainable agriculture. As Georgiana Kautz notes, the Nisqually Tribe's goal of improving tribal salmon and shellfish harvests will bring increased prosperity and quality of life to all residents of the watershed.

elementary school. Within three weeks she had recruited five additional and adjacent homeowners, and a noticeably positive and collaborative spirit emerged in her neighborhood.

Preparations for the May installation moved ahead at a fast pace, including individual rain garden construction sizing and design. Detailed event planning was also underway, with preparations for a weekend community planting event with volunteers and students and a live regional radio broadcast hosted by a well-known gardening personality, as well as collaborations with local environmental groups, businesses, the Nisqually Tribe, and the Town of Eatonville. Homeowners were enthusiastic, totally engaged, and helped each other throughout the project.

Site preparation and community planting events were completed during a single week, and the entire rain garden installation was hugely successful. The rain gardens were constructed during the week and ready for planting and mulching by volunteers at the Saturday planting events. Six rain gardens were installed, more than 60 volunteers were involved, and a group of neighbors that had never met or rarely spoken to each other before worked seamlessly toward a common goal. The positive outcomes greatly surpassed expectations, and this model continues to be used to install rain garden clusters in other neighborhoods in the Puget Sound area.

Key lessons from this project:

▶ **On a per garden basis, clustered rain garden construction is 30 to 35 percent cheaper than single project installations.**

▶ **Recruiting a neighborhood champion is the most important element in the process of educating and recruiting homeowners in a cluster.**

▶ **Success is inspiring. The community enthusiasm for these cluster installations generated additional funding to plan even larger cluster installations in Eatonville and Puyallup in 2011.**

▶ **Developing diverse partnerships between jurisdictions, tribal nations, local businesses, nonprofits, and interested citizens is essential in sustaining long-term behavior change.**

A completed rain garden in the Eatonville neighborhood.

Looking toward the old
Bowens Island restaurant
and an artificial oyster
reef along the shore.

179

SLIMY GOURMET: OYSTER RESTORATION IN THE CHARLESTON MARSHES

Around Charleston, South Carolina, tidal creeks flow up into cordgrass marshes. Bowens Island is a sandy, oak-covered rise that's above water at most tides. A restaurant of the same name on the island is famous for its beer and roasted oysters. The new LEED Silver–certified restaurant built after a 2006 fire destroyed the original kitchen features salvaged wood and traditional low-country design instead of air conditioning. As patrons walk up a ramp to the second-story restaurant, they get a close-up view of a living roof, built by landscaper Sam Gilpin on an adjacent building. Gilpin is in the process of converting an old cistern into a rainwater storage tank that will provide water for oyster washing and bathroom use. Eventually, the cistern will overflow into an extra-large rain garden, capable of soaking up the runoff from tropical storms, which can dump 8 inches (20 cm) in an hour.

Gilpin sees Bowens Island as a perfect opportunity to highlight the connection between urban runoff and the health of the Charleston harbor through the mollusk that is a regional staple, gracing high-end restaurant menus and backyard oyster roasts. Oysters and other shellfish are living water treatment plants, filtering nutrients and detritus from the waters around them. They are also sensitive to sediment and other pollutants in urban and agricultural runoff, and their numbers have declined in estuaries around the world in the wake of development. A museum under construction on the first story of the Bowens Island restaurant will showcase the history of the oyster industry, the environmental causes of its decline, and strategies for restoration.

One restoration project lies in the mud under the Bowens Island dock, which is usually lined with people fishing for whiting, flounder, red drum, and black sea bass. At the edge of the water oysters grow in rectangular mats on an engineered plastic substrate. These artificial reefs are increasingly deployed in areas where oysters once thrived and where the water is clean enough to support them. Oysters begin their lives as drifting plankton and then settle down on old oyster shells, which they use to attach their new shells to. In many estuaries, eroding soil from farms or clearcuts has buried old oyster reefs. Artificial reefs provide an attachment site for young oysters and eventually get completely covered with new shells. According to Charlestonian Emilie Agosto, locals prefer local oysters over others, hands down. This preference and the rise of the local food movement have been a boon for local oystermen. Charleston restaurants, long known for unique low-country cuisine, have recently focused on local meat, vegetables, and shellfish. One regional restaurant chain is starting an oyster farming operation to meet the demand, using artificial reefs similar to the ones at Bowen's Island. The local food movement may improve the region's oystering economy and water quality in the estuary.

URBAN COMMONS, DESERT ORCHARDS, AND FISH-FILLED GREENHOUSES

Many cities are transforming curb strips (the long, skinny area between the sidewalk and the curb) into block-long rain gardens that filter stormwater running down the street. These gardens often feature native plants and provide beautiful, drought-tolerant wildlife habitat. But Desert Harvesters, based in Tucson, Arizona, goes one better by planting desert trees like mesquite and ironwood that have delicious and nutritious seeds in curbside rain garden basins. Desert Harvesters cofounder Brad Lancaster sees street orchards as a new urban commons and a community food resource that can sustain Tucsonans in times of need. In a city where many residents migrated from cooler and wetter climes, the street orchard project serves the vital cultural function of helping transplants appreciate the beauty and diversity of desert plants by bringing the plants into their gardens and kitchens.

The idea of a commons may seem out of place in contemporary cities, but Desert Harvesters' curbside commons are part of a long tradition of turning public urban spaces to productive use. Cows grazed the Boston Common (now a city park) until 1830. During World War II, New Yorkers turned Tompkins Square into vegetable gardens and 1.4 million British families grew vegetables in allotment gardens. In recent years, landowners and volunteers have begun restoring forests and hedgerows in the London Green Belt, a large community forest on the outskirts of the city. In the 1970s, Melbourne gardeners transformed a dump into the Centre for Education and Research in Environmental Strategies, which now features edible and native gardens, a rainwater harvesting café, greywater treatment wetlands, and an array of rain gardens. Groups of gardeners in Philadelphia and Detroit have transformed vacant lots into social gathering places surrounded by fruit trees and flowers. Organizations such as Milwaukee's Growing Power and Detroit Summer have partnered with cities to build farms on city-owned lots and create new sustainable economies in the American Rust Belt, a region that once had a vibrant steel industry. Like Desert Harvesters, these organizations emphasize the cultural benefits of growing food locally: increased access to healthy food, an opportunity to share farming and cooking traditions, and a community spirit forged through collective work for the common good.

Growing Power's founder, Will Allen, is recognized internationally for using urban farms to recycle food waste into compost, improve food security in urban areas, and develop sustainable economies. To grow food throughout Milwaukee's frigid winters, Growing Power developed aquaponics, a system that uses recirculating water and biological filtration to grow fish and vegetables in greenhouses. An 18,000-gallon (68,500-L) rainwater harvesting system installed in 2010 harvests rain from the greenhouse roofs. Perch, trout, and tilapia swim in the cisterns and enrich the water with nitrates, which encourages plant growth. The nitrate-enriched water is then used to irrigate crops, and solar pumps return the plant-cleaned water to the cisterns. Rain tops up the tanks, replacing water that the plants transpire.

" LAST WORDS

THESE CASES illustrate creative extensions of rain gardening into larger collaborations with natural systems. The theme that runs through them all is sustainability and food. If people are going to inhabit a watershed, they need to get their food there. This is the message of the food justice movement—that all people deserve fresh, healthy food. Will Allen calls this movement the Good Food Revolution and emphasizes economics, as well as health and environmental welfare, as goals. In Milwaukee, Allen wants to revive African-American farming traditions and create good livelihoods for urban farmers. On the Nisqually, Billy Frank Jr.'s goal for the salmon recovery effort is to create a hundred permanent fishing jobs for tribal members and subsistence for many others.

Whether that food is salmon, oysters, perch, or mesquite, producing it requires collaboration with the rain through an adaptation of the rain garden concept to local climate and need. We hope your rain garden will inspire you to seek out similar collaborations in your neighborhood and along your local river, stream, or estuary. Whether you like to write letters or plant trees, paint pictures of spawning salmon or count them, mentor youth or organize senior citizens, your watershed needs you. Ongoing collaborations with the people, plants, rocks, animals, and water in your home landscape will expand your positive impact beyond your wildest dreams.

ECOLOGICAL
REGIONS

THE MESH AND INTERPLAY OF GEOLOGY, SOIL, LANDFORM, PLANTS, CLIMATE, WATER, AND HUMAN INTERACTIONS

gives rise to distinct ecological regions, also known as ecoregions. An ecoregion is larger than a biome (a deciduous woodland or cypress swamp, for example) and encompasses many watersheds and microclimates.

Following the World Wildlife Federation's classification system, we present five ecoregions that together encompass most of North America, the United Kingdom, Australia, and New Zealand. (For a map of the world's ecoregions, see wwf.panda.org/about_our_earth/ecoregions/about/habitat_types/selecting_terrestrial_ecoregions/). These ecoregions contain many smaller watersheds and biomes, but each is similar enough in climate and soil that gardeners from around the world who share the same ecoregion can grow many of the same plants. We hope this brief overview of the ecoregions in North America, Australia, and the United Kingdom will help you select plants from our lists and from other rain garden handbooks.

TEMPERATE CONIFEROUS FORESTS

In the United Kingdom, a vast woodland of Scots pine, birch, rowan, aspen, juniper, and oak—called the Caledonian Forest—once covered much of Scotland; only 1 percent remains today. In North America, coniferous forests stretch from California to Alaska, cover much of the Rocky Mountains, and blanket the southeastern U.S. lowlands from North Carolina to Louisiana. Along the Pacific Coast of North America, the cool wet climate and rich soils support forests of giant cedars, firs, and pines and feed large fish-filled rivers. Snow falls in the higher elevations of the coastal mountains, and summers are cool and dry. Stretching from interior Alaska through British Columbia and Alberta to New Mexico and eastern California, the taller Rocky Mountains, Sierra Nevada, and Cascade Mountains form the backbone of western North America. The forests in these ranges are hot in the summer, snowy in the winter, and rocky all the way. Rainfall increases to the north, from an average of 12 inches (30 cm) per year in New Mexico and the southern Sierra Nevada to 150 inches (375 cm) a year in the Olympic rain forest. South of Canada, summers are hot and dry, so plants need to tolerate extremes of temperature as well as moisture.

Conifers thrive in thin soils that are low in nutrients. In formerly glaciated regions of the United Kingdom and North America, soils may be poorly drained, challenging rain gardeners who build oversized basins and want to experiment with seasonally dry vernal pools.

TEMPERATE BROADLEAF AND MIXED FORESTS

In North America, this large ecological region extends from the Atlantic coast westward approximately 400 miles (650 km) into eastern Texas, Oklahoma, Missouri, Iowa, and Minnesota and encompasses large urban areas such as Boston and Washington, D.C., as well as those near the Great Lakes such as Chicago, Detroit, and Toronto. Although temperatures vary widely over the region, rainfall patterns are similar. Rainfall ranges from 40 to 70 inches (100 to 175 cm) per year, with most rain falling during the summer (the hurricane season in coastal areas). Abundant rainfall sustains dense deciduous forests, which create rich, organic soils. In the lowlands, pine forests and pocosin dominate dry sites, and cypress swamps and marshes gradually give way to open water.

In Australia, this ecoregion covers Tasmania; lines the coast of Victoria, New South Wales, and Queensland; and encompasses the Melbourne, Sydney, and Brisbane urban areas. *Eucalyptus* and *Acacia* dominate temperate broadleaf and mixed forests in Australia. The temperate forests stretching from southeastern Queensland to southern Australia enjoy a moderate climate and high rainfall that give rise to unique *Eucalyptus* forests and open woodlands. Northern New Zealand features temperate rain forests and highland forests.

In the United Kingdom, the English lowland beech forests once dominated the low-lying chalk lands of southeastern England, while Celtic broadleaf forests once covered the higher, drier metamorphic and igneous province of the northwest.

Despite this diversity of biomes, many rain garden plants will thrive equally well in New England and the Deep South, Auckland, and Essex. Although this ecoregion is generally humid, rain gardens in thin or well-drained soils should be designed to tolerate drought as well as flooding.

GRASSLANDS, SAVANNAS, AND SHRUBLANDS

In North America, prairies extend for about 950 miles (1500 km) from Alberta, Saskatchewan, and Manitoba, south through the Great Plains of the United States to southern Texas and adjacent Mexico, and approximately 600 km from western Indiana to the foothills of the Rockies and into northeastern Mexico. Rainfall decreases from more than 50 inches (125 cm) per year in eastern Oklahoma to less than 10 inches (25 cm) per year in western Texas.

Tropical grasslands stretch in a broad band along Australia's northern coast, yielding to temperate grasslands in interior Queensland and New South Wales. Grasslands also dominate New Zealand's southern coast.

This ecoregion is characterized by grasslands, thick topsoil, hot summers, and cold winters. Rain gardens thus need to tolerate extreme temperatures and frequent flooding. In drier areas, plants must also tolerate drought.

DESERTS AND XERIC SHRUBLANDS

In North America, deserts and xeric shrublands extend from eastern British Columbia in the north to Baja California and north-central Mexico in the south. This ecoregion lies in the rain shadow of the Cascade Range, Sierra Nevada, and Sierra Madre, and for most of the year, more water evaporates or transpires than falls. Temperatures vary from 110°F (43°C) during Sonoran Desert summers to –5°F (–20°C) during Great Basin winters. In Australia, this ecoregion dominates in the interior.

In most deserts, searing daytime heat gives way to cold nights because clouds rarely provide insulation. Although quite harsh, the diversity of climatic conditions supports a rich array of habitats. Many are ephemeral in nature, reflecting the seasonality of available water. Minerals in the soil inhibit drainage, so plant roots play a crucial role in increasing infiltration. Grasses and herbaceous perennials do best in the wet zone of a rain garden, whereas drought-tolerant shrubs and succulents thrive on the dry slopes and berm.

MEDITERRANEAN FORESTS, WOODLANDS, AND SCRUB

Mediterranean biomes feature uniquely adapted animal and plant species, which can handle the stressful conditions of long hot summers with little rain. Most plants are fire-dependent, meaning that they require regular fires in order to regenerate. In North America, this relatively small ecoregion extends from northern California to Baja California in the south. It abuts the Pacific Ocean on the west and the Sierra Nevada and deserts to the east. It is distinguished by its warm and mild climate and its shrubland vegetation of chaparral mixed with areas of grassland and open oak woodlands. Southwestern and south-central Australia's Mediterranean biomes feature extraordinary plant diversity, with more than 5500 species of shrubs, 70 percent of them endemic.

Mediterranean regions experience extremely variable precipitation, with rainfall ranging from 3 to 25 inches (7.5 to 62.5 cm) and varying from year to year. Rain falls only in the winter, and native plants grow during the rainy season and go dormant during the six-month dry season. Plants from other Mediterranean climates, including the Mediterranean basin, Chile, and South Africa, are good rain garden candidates for Mediterranean rain gardens in California and Australia. Plants from other regions will require summer irrigation.

RESOURCES

THIS RESOURCE LIST

contains those materials we have found to be the most helpful and inspiring. There are many more resources to be found, so please continue beyond this list and find what will inspire you most.

BOOKS AND FILMS

Agarwal, Anil, and Sunita Narain. 1997. **DYING WISDOM: RISE, FALL AND PO-TENTIAL OF INDIA'S TRADITIONAL WATER HARVESTING SYSTEMS.** Centre for Science and Environment, New Delhi, India.
A fascinating survey of traditional rainwater harvesting systems across India's diverse climates and landscapes.

Dunnett, Nigel, and Andy Clayden. 2007. **RAIN GARDENS: MANAGING WATER SUSTAINABLY IN THE GARDEN AND DESIGNED LANDSCAPE.** Timber Press, Portland, Ore.
Written for landscape architects, garden designers, and gardeners, this book provides a history of low-impact development and a great introduction to the water cycle. It features examples of rain gardens in housing developments and public places.

Dunnett, Nigel, Dusty Gedge, John Little, and Edmund C. Snodgrass. 2011. **SMALL GREEN ROOFS: LOW-TECH OPTIONS FOR GREENER LIVING.** Timber Press, Portland, Ore.
This book showcases small-scale living roofs throughout the world, including roofs on sheds, garden offices, studios, garages, and bicycle sheds, as well as several community projects. The authors describe design, construction, installation, and how each roof was planted and cared for.

Kinkade-Levario, Heather. 2007. **DESIGN FOR WATER: RAINWATER HARVEST-ING, STORMWATER CATCHMENT, AND ALTERNATE WATER REUSE.** New Society Publishers, Gabriola Island, B.C.
This somewhat technical book provides a good overview of rainwater harvesting in cisterns and discusses how to integrate high-tech greywater systems and stormwater catchment.

Lancaster, Brad. 2006. **RAINWATER HARVESTING FOR DRYLANDS, VOL. I: GUIDING PRINCIPLES TO WELCOME RAIN INTO YOUR LIFE AND LAND-SCAPE.** Rainsource Press, Tucson, Ariz.
Lancaster, Brad. 2008. **RAINWATER HARVESTING FOR DRYLANDS AND BEYOND, VOL. 2: WATER-HARVESTING EARTHWORKS.** Rainsource Press, Tucson, Ariz.
These thorough and accessible guides cover all aspects of rainwater harvesting and are chock full of humor and useful water information. Highly recommended for rain gardeners who want to experiment with integrative design. Available at www.harvestingrainwater.com.

Ludwig, Art. 2005. **WATER STORAGE: TANKS, CISTERNS, AQUIFERS, AND PONDS FOR DOMESTIC SUPPLY, FIRE, AND EMERGENCY USE.** Oasis Design.
This self-published guide explains how to build cisterns and ponds and includes detailed plans for ferrocement cisterns. Ludwig's *Create an Oasis with Greywater* is a great source for how-to greywater information. Available at www.oasisde-sign.net.

Pepin Silva, Elizabeth. 2010. **"SLOW THE FLOW: MAKE YOUR LANDSCAPE ACT MORE LIKE A SPONGE."**
This 30-minute film brings to life projects that individuals and communities have created to slow down the flow of stormwater. The projects and approaches highlighted are very low-tech, inexpensive, and beautiful—making a good argument for kicking back and not raking the leaves or watering the lawn. Available at http://vimeo.com/14722643.

Reynolds, Michael. 2005. **WATER FROM THE SKY.** Solar Survival Press, Taos, N.M.
A primer for water harvesting in off-grid earthship homes, this book is an inspiring read regardless of what type of home you live in. It includes a broad perspective on water issues, thorough consider-

ations for the design of a water harvesting system, and many creative solutions and strategies based on Reynolds' years of hands-on experience.

Schwenk, Theodor. 1965. **SENSITIVE CHAOS: THE CREATION OF FLOWING FORMS IN WATER AND AIR.** Rudolph Steiner Press, London.
A classic book that poetically describes the flow patterns of water in nature. Schwenk simplifies reoccurring patterns and rhythms found throughout all living ecosystems, whether it be drops of rain hitting a surface or human cells growing.

Wilkinson, Charles F. 2006. **MESSAGES FROM FRANKS LANDING: A STORY OF SALMON, TREATIES, AND THE INDIAN WAY.** University of Washington Press, Seattle.
This oral history covers the fish wars and Billy Frank Jr.'s three decades of work to restore the Nisqually River. A great read for anyone interested in tribal sovereignty and ecological restoration.

Woelfle-Erskine, Cleo, July Oskar Cole, Laura Allen, and Annie Danger. 2007. **DAM NATION: DISPATCHES FROM THE WATER UNDERGROUND.** Soft Skull Press, New York, N.Y.
An overview of water history in the United States and current water battles around the world, as well as sustainable alternatives. It includes practical information on rainwater harvesting, greywater reuse, and composting toilets

Woelfle-Erskine, Cleo. 2003. **URBAN WILDS: GARDENERS' STORIES OF THE STRUGGLE FOR LAND AND JUSTICE.** water/under/ground publications, Oakland, Calif.
Cleo's early writing on urban gardening, with an urban permaculture slant.

RAIN GARDEN GUIDES

Every year there are more countries, states, and provinces advocating rain gardens, and with this comes more locally adapted rain garden resources. If you can't find one for your area, contact local agencies and encourage them to become active with rain gardens, or start your own community group to educate the public and install rain gardens in your area.

AUSTRALIA
MELBOURNE WATER'S 10,000 RAIN-GARDEN PROGRAM: http://raingardens.melbournewater.com.au

NORTH AMERICA
ANCHORAGE, ALASKA, RAIN GARDENS: http://www.anchorageraingardens.com

ARIZONA RAIN GARDEN GUIDE: HARVESTING RAINWATER FOR LANDSCAPE USE: http://cals.arizona.edu/pubs/water/az1344.pdf

GREEN VENTURE WISE WATER USE: http://water.greenventure.ca/rain-gardens (Canadian focus)

INDIANA: A RAIN GARDEN HOW-TO MANUAL FOR CITY OF FORT WAYNE HOMEOWNERS: http://www.catchingrainfw.org

IOWA RAIN GARDEN DESIGN AND INSTALLATION MANUAL: http://www.ia.nrcs.usda.gov/features/raingardens.html

OREGON RAIN GARDEN GUIDE: LANDSCAPING FOR CLEAN WATER AND HEALTHY STREAMS: http://seagrant.oregonstate.edu/sgpubs/onlinepubs.html
PRINCE GEORGE'S COUNTY, MARYLAND, BIORETENTION MANUAL: http://www.princegeorgescountymd.gov

RAIN GARDEN NETWORK: http://www.raingardennetwork.com

SOUTH CAROLINA RAIN GARDEN MANUAL: http://media.clemson.edu/public/restoration/carolina%20clear/toolbox/publication_raingardenmanual_022709b.pdf

TEXAS MANUAL ON RAINWATER HARVESTING: http://www.twdb.state.tx.us/publications/reports/rainwaterharvestingmanual_3rdedition.pdf

TORONTO HOMEOWNER'S GUIDE TO RAINFALL: http://www.riversides.org/rainguide/index.php

TUCSON, ARIZONA, RAINWATER HARVESTING MANUAL: http://dot.ci.tucson.az.us/stormwater/downloads/2006WaterHarvesting.pdf

WISCONSIN RAIN GARDENS: A HOW-TO MANUAL FOR HOME OWNERS: http://dnr.wi.gov/runoff/rg

UNITED KINGDOM WILDFOWL AND WETLANDS TRUST: http://www.org.uk/our-work/wetland-habitats/rain-gardening

PLANTS

Local nurseries are great resources to find plants suitable for your rain garden. There are too many to list here, so scout out some near you and ask the staff about water-loving species for your region. Rain garden guides, like those listed above, also have good plant information. Below are a few online sources that have excellent databases of information on rain garden plants.

10,000 RAIN GARDENS: www.rainkc.com (North America)

AUSTRALIAN NATIONAL BOTANICAL GARDEN: http://anbg.gov.au/ (Australia)

AUSTRALIAN NATIVE PLANT SOCIETY: http://anpsa.org.au/index.html (Australia)

BRITISH WILDFLOWER PLANTS: http://www.wildflowers.co.uk/ (Europe)

LINGFIELD RESERVES (suitable species listed in Damp Ground and Pond Margin): http://www.lingfieldreserves.org.uk/wetland_plants.htm (Europe)

NATIVE RAIN GARDEN: http://www.native-raingarden.com (North America, Europe, and Australia)

RAIN SCAPING: http://www.rainscaping.org (North America)

THE ROYAL TASMANIAN BOTANICAL GARDEN: http://www.rtbg.tas.gov.au/file.aspx?id=553 (Australia)

THREE RIVERS RAIN GARDEN ALLIANCE: raingardenalliance.org (North America)

TOHONO CHUL PARK (DESERT PLANTS): www.tohonochulpark.org (North America)

WATER POLICY AND EDUCATION

BRAD LANCASTER (www.harvestingrainwater.com) shares useful technical information, photos, videos, plant lists, and an up-to-date listing of water projects and suppliers, mostly in the American Southwest, but with some international links.

GREYWATER ACTION (www.greywateraction.org) features photographs and plans for rainwater harvesting, greywater, and composting toilet systems, as well and policy alerts and a list of workshops. California and U.S. focus, with an emphasis on urban systems.

HARVEST H2O (http://harvesth2o.com) includes links to current water news, case studies and research, how-to instructions, community blog, and other resources.

OCCIDENTAL ARTS AND ECOLOGY CENTER (http://oaecwater.org) produces handbooks and reports on rainwater harvesting and watershed literacy, with a rural, Mediterranean climate focus.

THE COMMUNITY COLLABORATIVE RAIN, HAIL, AND SNOW NETWORK (www.cocorahs.org) is a group of volunteers working together to measure precipitation across the nation. If you become a member, you receive free training on how to check your rain gauge, then keep track of daily totals and report them to the network. Look on their website for volunteers in your area—you can use their rainfall data as you design your rain garden.

CASE STUDIES AND FORUMS

Here is a short list of organizations that put rain gardens to work in their local areas. Check out some of these case studies for inspiration in your own garden.

CENTRE FOR EDUCATION AND RESEARCH IN ENVIRONMENTAL STRATEGIES, a community eco-park in Melbourne that features rainwater harvesting and greywater systems, among other sustainable systems: http://www.ceres.org.au/greentech/water (Australia)

GREEN INFRASTRUCTURE WIKI: http://www.greeninfrastructurewiki.com/page/Case+Studies%3A+Planning+and+Policy (North America and the United Kingdom)

KINGSTON, VICTORIA, RAIN GARDEN SITE: http://www.kingston.vic.gov.au/Page/page.asp?Page_Id=1429 (Australia)

LAKE SUPERIOR STREAMS: http://www.lakesuperiorstreams.org/stormwater/toolkit/raingarden.html (North America

RAIN GARDEN NETWORK: http://www.raingardennetwork.com/photos.htm (North America)

STRINGYBARK CREEK RAIN GARDEN PROJECT in Mount Evelyn, Victoria, part of a research program coordinated by the University of Melbourne and Monash University (videos and case studies): http://www.urbanstreams.unimelb.edu.au (Australia)

SUSTAINABLE GARDENING AUSTRALIA: http://www.sgaonline.org.au (Australia)

WATER BY DESIGN, showing water-sensitive urban design projects in the Brisbane area: http://waterbydesign.com.au/case-studies (Australia)

WATER-SENSITIVE URBAN DESIGN, showing projects in the Melbourne area: http://wsud.melbournewater.com.au/content/case_studies/case_studies.asp (Australia)

PHOTO
CREDITS

INDEX

ABOUT THE AUTHORS

CLEO WOELFLE-ERSKINE is a hydrologist, educator, and scholar of water. He cofounded the Greywater Guerrillas and has led dozens of community-based workshops on rainwater harvesting and greywater reuse across North America. His writing on dam removal, environmental justice, and urban gardening has appeared in *High Country News* and *YES!*, among other magazines, and in his anthologies *Urban Wilds: Gardeners' Stories of the Struggle for Land and Justice* and *Dam Nation: Dispatches for the Water Underground*. An avid gardener, Cleo is also pursuing a Ph.D. in the Energy and Resources group at the University of California–Berkeley, where he is investigating how rainwater harvesting affects stream flow and how rainwater harvesting helps cities cope with climate change.

APRYL UNCAPHER is an ecological designer dedicated to rainwater harvesting modeled on nature's examples. Coauthor of the *San Francisco Rainwater Harvesting Guide*, Apryl has consulted on integrative stormwater management and water recycling projects for more than a decade. With a degree in Design Engineering from Purdue University and accredited as a LEED professional, Apryl continually seeks opportunities to participate in water projects that inspire a sustainable water consciousness. At her home in San Francisco, Apryl passionately preserves water resources with two little girls splashing and dreaming by her side.